A perspective of
environmental pollution

M. W. HOLDGATE

CAMBRIDGE UNIVERSITY PRESS

Cambridge
London   New York   Melbourne

Published by the Syndics of the Cambridge University Press
The Pitt Building, Trumpington Street, Cambridge CB2 1RP
Bentley House, 200 Euston Road, London NW1 2DB
32 East 57th Street, New York, NY 10022, USA
296 Beaconsfield Parade, Middle Park, Melbourne 3206, Australia

First published 1979

Printed in Great Britain at the
University Press, Cambridge

Library of Congress Cataloguing in Publication Data
Holdgate, Martin W.
A perspective of environmental pollution.
Bibliography: p.
Includes index.
1. Pollution. 2. Ecology. I. Title.
TD174.H63   363.6   78-8394
ISBN 0 521 22197 8

Contents

Preface   *page* viii
Acknowledgements   x

1 The ecological context   1
   1.1   The two worlds of man   1
   1.2   The ecological background   2
         1.2.1   Biogeochemical cycles   2
         1.2.2   Ecosystems   3
         1.2.3   Succession   8
         1.2.4   Life strategies   9
         1.2.5   Stress and resilience   11
   1.3   The impact of man   12
         1.3.1   The stress of pre-industrial man   12
         1.3.2   The development of agriculture   13
         1.3.3   Industrial stress   15
   1.4   The context of pollution   16

2 Pollution and pollutants   17
   2.1   Definitions   17
   2.2   Pollution by energy and sound   18
   2.3   Pollutant substances   19
   2.4   Classification of pollutants   22
   2.5   Pollutants in environmental sectors   25
         2.5.1   Pollutants in the air   25
         2.5.2   Pollutants of water   27
         2.5.3   Pollutants of food   31
         2.5.4   Pollutant substances: consensus of views   31
   2.6   Properties of pollutant substances   35
   2.7   Prediction of pollution hazard   40
         2.7.1   Pointers to a problem   40
         2.7.2   Screening   41
         2.7.3   National and international data banks   42

3 The significance of pathways   44
   3.1   Terminology: sources, pathways and sinks   44

Contents

3.2 Pathways in the environment   45
   3.2.1 Pathways in air   46
   3.2.2 Pathways in water   52
   3.2.3 Pathways in soil   55
   3.2.4 Pathways modified by man   55
   3.2.5 Interlinkages between pathways   56
3.3 Pathways within targets   57
   3.3.1 Pathways in ecosystems   57
   3.3.2 Pathways within organisms   57
   3.3.3 Pathways in food chains   62

4 Changes in the environment   64
4.1 The significance of environmental concentrations   64
4.2 Changes in the media   66
4.3 Physical effects   66
   4.3.1 Heating the air   66
   4.3.2 Heating the waters   69
   4.3.3 The transmission of radiation through the atmosphere: possible effects of carbon dioxide   69
   4.3.4 Transmission of radiation: possible effects of aircraft, chlorofluorocarbons and oxides of nitrogen   74
4.4 Chemical changes in the air: trends in other atmospheric contaminants   78
   4.4.1 Sulphur dioxide and smoke   78
   4.4.2 Gaseous emissions from vehicles   83
   4.4.3 Emissions of metals, droplets and particles   84
4.5 Changes in rivers, lakes and estuaries   88
   4.5.1 'Eutrophication' and sewage   88
   4.5.2 Overall trends in fresh-water quality   91
4.6 Changes in the sea   97
   4.6.1 Oil at the air–water interface   97
   4.6.2 Changes in the chemical composition of the seas   99
   4.6.3 Changes in British inshore seas   101
4.7 Changes on land   103
4.8 Radioactivity in the sea and on land   105
4.9 The scale of change   105

5 The effects of pollution on targets   108
5.1 Effects and targets   108

5.2   Effects of sound, radiation and temperature on living targets   116

5.3   Effects of chemical pollutants on animals   118

     5.3.1   Biochemical and physiological effects   118

     5.3.2   Physiological responses   121

     5.3.3   Carcinogenesis, mutagenesis and teratogenesis   124

     5.3.4   Effects on animal populations   125

5.4   Effects on plants   129

     5.4.1   The nature of pollutant effects   130

     5.4.2   Interactions between pollutants, and between pollutants and other variables   130

     5.4.3   Variation between species and taxonomic groups   132

     5.4.4   Patterns of plant performance in nature   132

5.5   The ecosystem level   134

6   Target exposure, risk and the establishment of goals and standards   140

6.1   Risk estimation   140

6.2   Standards and objectives   143

     6.2.1   Primary protection standards   144

     6.2.2   Environmental standards   148

     6.2.3   Emission standards   155

     6.2.4   Other types of standard   160

6.3   The inevitability of change and variation in standards   162

7   Monitoring and surveillance   164

7.1   The purpose of monitoring   164

7.2   The design of a monitoring scheme   166

     7.2.1   What to measure   166

     7.2.2   Where and when to make measurements   170

     7.2.3   Some characteristics of factor monitoring schemes   172

     7.2.4   Some characteristics of target monitoring schemes   173

7.3   Monitoring in operation   178

     7.3.1   Global, regional and national surveillance and monitoring   178

     7.3.2   Monitoring in Britain   180

7.4   The components of a monitoring scheme   189

8   Costs and controls   190
    8.1   Basic questions   190
    8.2   The costs of damage and the evaluation of benefits   191
          8.2.1   Damage costs   192
          8.2.2   Environmental values   195
          8.2.3   Benefits   196
          8.2.4   Control costs   197
          8.2.5   Financing controls   199
    8.3   Mechanisms for the control of pollution   201
    8.4   Planning and assessment of the environmental
          impact of new development   202
          8.4.1   The concept of environmental impact
                  assessment   202
          8.4.2   The approach to environmental impact
                  assessment   203
          8.4.3   Pollution avoidance through develop-
                  ment control   206
    8.5   Regulation of pollution   211
          8.5.1   Central strategy and co-ordination   211
          8.5.2   Decentralisation to local government or
                  specialist agencies   213
          8.5.3   The overall pattern   217

9   Pollution as an international problem   220
    9.1   The nature of the problem   220
          9.1.1   Transfrontier pollution   220
          9.1.2   Pollution from mobile sources and
                  products   221
          9.1.3   Effects on the world environment   222
          9.1.4   Effects on trade and economic balance   222
    9.2   The scale of the problem   223
          9.2.1   National and regional 'hot spots'   223
    9.3   Questions to be answered   225
          9.3.1   Kinds of question   225
          9.3.2   Scientific questions: research, monitor-
                  ing and evaluation   226
          9.3.3.   Questions relating to the use of the
                  environment as a sink   227
          9.3.4   Questions of sources and targets   230
          9.3.5   Standards   231
          9.3.6   Questions of harmonising enforcement
                  policies   233

9.4   International action to control pollution   235
      9.4.1   Laws and conventions   235
9.5   Conclusions: the pattern of international action   238

10   The approach to the future   241
     10.1   Is pollution a threat?   241
             10.1.1   Human population growth and its
                       limitation   241
             10.1.2   Pollution and mortality in man   244
             10.1.3   The establishment of safety margins 245
             10.1.4   The capacity for pollution control 247
             10.1.5   The threat from pollution: a ten-
                       tative conclusion   248
     10.2   Future trends in our approach to pollution   249
     10.3   The approach to environmental management   252
     10.4   A final summary   255

References   258
Index   271

Preface

In the past ten years there have been many books and reports on the environment. They vary enormously in scope, scientific content and tone. Some treat the changes in the world as a whole, especially those resulting from expanding industry, massive increases in the generation of energy, spreading cities and the rapid population growth of countries in the Third World. Still more focus on particular regions or on topics such as air or water pollution, the impact of persistent pesticides on wildlife, or the legal and administrative measures different countries have adopted in response. Some unashamedly set out to stimulate public concern and so provoke more stringent control of pollution and other kinds of 'environmental impact'. Others confine themselves to the scrutiny of scientific evidence about causes and effects or to the technology which may allow us to retain the benefits of industrial processes without their unwelcome side-effects.

The present volume takes a different line. It seeks to examine the principles which lead us to consider a substance as a pollutant, and govern the way we then respond. For this reason, the book covers a wide conceptual field, from the ecological context within which pollutants act through the nature, sources, pathways and effects of pollutant substances to the social elements of our response: monitoring, standards, controls and international conventions. The detailed evidence has been selected to provide examples rather than a complete coverage, and I am conscious of the fact that by the date of publication new knowledge will have qualified my interpretation at many points.

None the less, I believe that the central thesis will stand. It is that we must judge pollution in context. What we call pollution is the perturbation by man of a series of physical and chemical factors and cycles, with consequent effects on living and non-living systems. The scale and nature of this perturbation needs to be measured so that we know more clearly what we are doing to ourselves and to the world, and can relate the action we take realistically to the scale of the problem and to the risks.

This need to balance risk and response – and the uncertainties we at present face in trying to do so – runs through these pages. It is important, because our societies are still founded on both the natural and the man-made world. We depend on the former for the basic necessities of air, food and water, and for the less tangible (but still universally sought) benefits of natural beauty. From the man-made world, built through generations of inventiveness, skill and effort, we have gained some protection from the vagaries of natural forces and the freedom to create ideas, arts – and books. The debate about 'the environment' is concerned with the interaction between these two divisions of our world. At this level of generality there is no single, simple diagnosis of the health of that relationship – or cure for symptoms of ill. Because the world is environmentally and economically diverse, problems are rarely universal. Total gloom or total euphoria are rarely appropriate emotions, and societies will rightly differ in the detail of their priorities. A critical approach to the definition of the scale and nature of problems, and to the development of an appropriate response, may however be a universal need. This book gives a perspective – no doubt one among many alternatives–of such an approach. It is aimed at serious students of the environment – scientists or administrators – trying to place their own specialist knowledge in a wider context or grasp the shape of the overall pattern of concern.

There is an element of circularity in the argument developed in this book. Chapter 2 talks about the substances that may be pollutants, and how to classify them – but it is the changes a substance brings about in the environment and its effects on targets, discussed in Chapters 4 and 5, that make it a pollutant. The appropriateness of a particular classification depends on circumstances: for example on whether we are using it to place such effects in some kind of order, considering the acceptability of risk (Chapter 6) or concerning ourselves with controls (Chapter 8). Hence no particular classification can be justified without knowing what it is being used for. Similarly, the establishment of standards and objectives, discussed in Chapter 6, must not only be related to the damage a pollutant causes but must also take account of the cost of that damage and the feasibility of control, considered in Chapter 8. While the present 'perspective' offers one convenient and logical way through a complicated field, other perspectives will suit other purposes, and none will be uniquely valid.

I wrote these pages while I was Director of the Institute of Terrestrial Ecology, a component of the Natural Environment Research Council, between April 1974 and June 1976 (although I have made some alterations since then). My first-hand experience of pollution problems stems, however, from an earlier period as the first Director of the Central Unit on Environmental Pollution in the Department of the Environment. In this capacity I was privileged to work alongside colleagues who were endeavouring to improve the quality of the British environment and, through international negotiations, to safeguard the airs, lands and waters of the world. The views expressed are my own: and are not necessarily those of NERC or of the Department of the Environment by whom I was seconded to the Research Council. They should not be assumed to commit my successor Directors either in ITE or the Central Unit!

I would like to thank the colleagues and friends who read the book in draft and have improved it by their criticisms. In particular, I am grateful to Lord Ashby, FRS, and Professor J. W. L. Beament, FRS, for early encouragement and to Dr J. E. Peachey for comments on the later text. Dr P. J. W. Saunders gave me especial help by his thorough and critical attention to detail. My warm thanks are also due to Mr R. J. H. Beverton, FRS, Secretary of NERC, the present Director of ITE, Mr J. N. R. Jeffers, and to my secretaries at ITE, Miss M. J. Moxham, Mrs M. E. Chambers and Mrs J. Bowen, on whom the burden of production of the initial texts fell. Finally, I am deeply grateful to my family, on whose leisure this book has trespassed over-much.

M. W. HOLDGATE

Cambridge
December 1977

Acknowledgements

We are grateful to the Controller of Her Majesty's Stationery Office for permission to reproduce figures 22, 23, 24, 27, 28, 29, 30 and 32.

1. The ecological context

1.1 *The two worlds of man*

Man inhabits two worlds. One is the natural world of plants and animals, of soils and airs and waters which preceded him by billions of years and of which he is a part. The other is the world of social institutions and artefacts he builds for himself, using his tools and engines, his science and his dreams to fashion an environment obedient to human purpose and direction.

In this, the opening paragraph of their book *Only one earth*, written as a global perspective for the United Nations Conference on the Human Environment, held at Stockholm in 1972, Ward and Dubos (1972) sum up the basic essential of the human situation. Man is an animal with over three thousand million years of evolution behind him. That pathway of organic development has been shaped by interactions with other components of the natural world. Today we still depend on that world for our survival – for breathable air, drinkable water, food, and behind all, for the sun's energy which drives the whole complex system of living things. Ecology is the scientific study of living creatures in relation to their environment. It is the study of the first world of man.

But the outstanding feature of the last ten thousand years has been the construction of the second human world: the emergence of an increasingly complex social system, and the development of increasingly advanced skills whereby man has escaped, at least partly, from the vagaries of his natural background. Through forest clearance, the creation of pasture, the expansion of intensive cultivation, the alteration of drainage systems and the creation of technology man has increasingly changed the pattern of productivity of the natural world. This is development. He has been able, as a consequence, to increase his numbers greatly. As a by-product of the increase in his numbers and his technology, he has also greatly increased the volume and range of materials he has disposed of – often without much thought of the consequences – in the environment. Some of these wastes cause harm to the functioning of ecological systems and to man's own interests. This is pollution, and Chapter 2 describes various ways of classifying and evaluating the many substances that can cause it.

Development remains essential in many parts of the world where millions of people are inadequately fed and housed and lack satisfactory medicine, education and employment. But this development must be ecologically sound, and avoid the dissipation of natural resources through erosion, salinisation of badly irrigated soils, reduced productivity and pollution. The theme has been repeated in numerous reports in recent years (e.g. SCEP, 1970; Institute of Ecology, 1971; Kay and Skolnikoff, 1972; SCOPE/UNEP, 1974; Chadwick and Goodman, 1975; Institute for Cultural Research, 1975; UNEP, 1976, 1977; SCOPE, 1977). The term 'ecodevelopment' has been coined and used widely in United Nations circles. But it remains true that the objective of ecologically sound development is much easier to state than to achieve. This chapter examines some of the essential features of ecological systems which govern their response to pollution and need to be borne in mind when development and pollution control are planned, and Chapter 10 returns to this theme in a 'forward look' to the future.

1.2 *The ecological background*

1.2.1 *Biogeochemical cycles*

The earth has had three atmospheres. It had too small a mass to retain the first, made up of light gases and present at the time when the planet formed by the accretion of cold materials. The second was formed by the emission of gases like water vapour, methane, carbon dioxide and ammonia as the planet was heated by gravitational collapse and radioactive decay. The condensation of much of the water vapour subsequently formed the primaeval lakes and seas, and it was within this hydrosphere that the first life forms are thought to have evolved (Ponnemperuma, 1972, SCOPE, 1977). About two thousand million years ago a revolutionary change occurred, when the process of photosynthesis evolved, and with it the first plants. In this process, solar energy is used to power the combination of carbon dioxide and water to form carbohydrate, with the release of free oxygen into the air. The carbohydrate is subsequently used as a chemical energy source in respiration, atmospheric oxygen being recombined with carbon and carbon dioxide released. Today, all animals (including man) remain totally dependent on the photosynthetic activity of green plants, and it is this activity that has created the earth's third atmosphere with its abundant free nitrogen and oxygen, small quantities of water vapour and carbon dioxide, and very tiny traces of methane and ammonia.

There must have been geochemical cycles on the early earth. Elements passed from the rocks to solution in rivers and seas, were re-deposited in sediments, and might be mobilised again by further erosion and solution. Once plants became abundant, and increasingly complex ecological systems evolved, these cycles were changed and became biogeochemical cycles. Those of carbon, oxygen, nitrogen, phosphorus, sulphur and many metals are today largely regulated by living organisms (Kovda, 1975;

Svensson and Soderlund, 1976; SCOPE, 1977; SCOPE 13, in preparation). It is the perturbation of these cycles by the chemicals released by man and his technology (discussed in Chapter 4), and the changes consequent upon the alteration of vegetation and fauna (Chapter 5), that constitute the major problems of pollution.

1.2.2 *Ecosystems*

In the natural world, living things evolved within ecosystems. An ecosystem was defined by Tansley (1935) as 'a unit of vegetation which...includes not only the plants of which it is composed but the animals habitually associated with them, and also all the physical and chemical components of the immediate environment or habitat which together form a recognizable self-contained entity'. The core of this concept is the inclusion of physical and biological components in the system, and the constraint that the majority of the interactions occur within it – giving it the 'recognizable self-containment' and defining the frontier between adjacent ecosystems. This does not mean that all interactions must be internal to a properly defined ecosystem: a migratory bird, for example, may form part of different ecosystems at different seasons and an insect likewise inhabit different ones in its larval and adult stages. What it does mean is that ecological interactions tend to be localised. The biosphere is a mosaic of ecosystems and much of the detail of its functioning is lost if too coarse a scale of description – at global or regional level for example – is adopted. This is one reason why the action of pollutants needs to be examined especially at the local scale, within individual ecosystems, if the patterns of variation and the true scale of effects are to be defined.

The evolution of life, under the governing influences of genetic variation and natural selection, has produced species adapted to the diverse habitats of the world. The biomes, or biological zones, we recognise reflect this fact: it is a matter of common knowledge that the high-polar tundra and the high mountains are lands of lichen, moss and hardy herbaceous plants, while coniferous forests, deciduous forests, grasslands, steppes, arid deserts and tropical rain forests have each their proper region (see, for example, Huxley, 1973). The species present in these regions often have different ancestries, reflecting the way oceans and mountains have formed barriers to plant and animal migration in the past (for biogeographical details see, for example, Good, 1953; Darlington, 1957; Valentine, 1972), but display comparable adaptations both to physical features of the environment they live in and to competition with other species. Many authors have listed the principal physico-chemical factors that can be of ecological significance: Table 1 is based on the lengthy analysis by Allee, Emerson, Park, Park and Schmidt (1949).

Table 1. *Major physico-chemical factors affecting organisms*

Main factor	Form	Biological response
Physical factors		
Radiation	Light	Photosynthesis
		Photoperiodicity (seasonal and diurnal rhythms)
		Avoidance of damage from ultra-violet
		Adaptive coloration
		Visual adaptation
	Heat (direct and via medium, e.g. in stratified waters)	Rates of processes
		Thermoregulatory adaptations (including behaviour)
		Adaptations to extreme cold (hibernation, anti-freezes, tolerance of freezing, life cycle adjustment)
		Adaptations to extreme heat (shelter-seeking, aestivation, transpiration, life cycle adjustment)
Gravity, pressure and sound	Gravity	Structure (skeletal, growth form)
		Orientation
	Pressure	Tolerances and density adjustments (e.g. in deep-sea species)
	Sound and vibrations	Auditory and echolocatory adaptations
		Behavioural responses (avoidance of excess sound)
Currents of air and water	Wind	Dispersion: flight adaptations
		Resistance: skeletal and growth form adaptations, attachments to substratum, behavioural adaptations
	Ocean and fresh-water currents	Dispersion (various adaptations)
		Resistance (various attachments and behavioural adaptations, shelter seeking)
		Filter feeding adaptations in sedentary species
Substratum	Water surface	Physico-chemical adaptations to life on or under surface film (water birds, insects, water plants)
	Sea, lake or river bed	Adaptations for attached, rooted or burrowing life
	Soil and rock surface	depending on substratum type

Table 1 (*cont.*)

Main factor	Form	Biological response
Chemical factors		
Physico-chemical	Viscosity	Form adaptations in aquatic life (streamlining, buoyancy)
	Diffusion	Rates of uptake of component or dissolved materials in still air or water
	Osmosis etc.	Ionic uptake and body fluid regulation in aquatic species
	pH	Influences wide range of species and processes: adaptations to particular acidity or alkalinity levels in water and soil solutions
	Adsorption	Adsorption of dusts on surfaces affects permeability (waterproofing). Effects on ionic behaviour and availability in soil
Water composition and form	Physical state (ice, snow, humidity)	Adaptations to sustain water uptake and prevent excessive loss: waterproofing Temperature and water interactions often critical for species
	Dissolved salts	Major limiting factor for many species (especially nitrogen, phosphorus, calcium, magnesium, silicon) Adaptations for absorption even from dilute solution, and retention against concentration gradient
	Dissolved gases (especially oxygen and carbon dioxide)	Adaptations in aquatic life to secure adequate supplies: wide variation in tolerance of anoxia
Atmospheric composition	Gases	Oxygen and carbon dioxide essential to life: other gases e.g. carbon monoxide sulphur oxides, ozone may damage sensitive tissues and surfaces
	Suspended particles	Various adaptations against excessive deposition

After Allee, Emerson, Park, Park and Schmidt (1949).

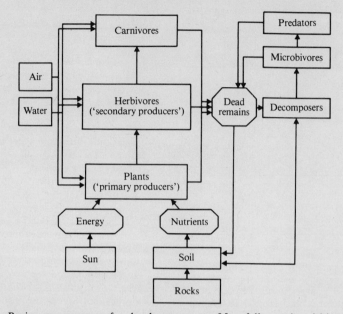

Figure 1. Basic components of a land ecosystem. Man falls partly within the herbivore, and partly within the carnivore box.

Plant and animal species alike have evolved to flourish best within certain extremes of temperature. Levels of dissolved oxygen are critical in determining the fish and invertebrate fauna of lakes and rivers. Trout do not survive in lakes more acid than around pH 4.0 Shellfish require a certain availability of calcium salts if they are to survive and grow. Different species of aquatic and terrestrial plants have different demands for nutrients like nitrogen and phosphorus. The tolerances of species are very different, which is one reason why it is hard to predict all the effects of any particular chemical pollutant without thorough physiological and ecological study of the organisms concerned (see Chapter 5).

Broadly, ecosystems contain five main functional categories of component organism (Figure 1). At the base, driven by the input of nutrients, carbon dioxide and solar energy are the green plants or primary producers, using the sun's energy to make carbohydrates. On them feed herbivorous animals, or secondary producers. In turn these sustain carnivores, forming a tertiary 'trophic level'. Sometimes the chain is longer, as when an insect-eating bird is the prey of a larger predator like a hawk.

But the flow of energy through an ecosystem should not be thought of as simply following a straight line from plant to herbivore to carnivore. Much plant material is deposited, along with animal body wastes and carcases, as dead matter on the ground or the bed of a water body. Here

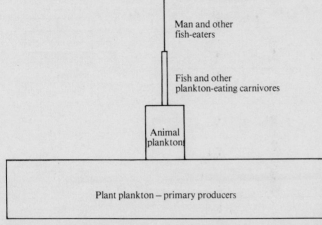

Figure 2. 'Pyramid' of mass of living organisms in the trophic levels of an ecological system in the sea.

it sustains a major component of the ecosystem, the decomposers: chiefly bacteria and fungi. In turn these are consumed by microbivores, which may themselves sustain higher predators. All these components play their part in the biogeochemical cycles, and when their inter-relationships are disturbed by pollution, for example when excessive input of nutrients causes the phenomenon of eutrophication of lakes described in Chapter 4, major local changes in the cycles result.

Far more solar energy falls on the earth's surface than is fixed by green plants. Calculations vary, but Nobel (1973) suggested that around 5% of the solar radiant energy reaching the earth was intercepted by green plant surfaces, and of this only some 1% was converted into the chemical energy stored as carbohydrate. The efficiency of conversion of sunlight into plant food produce used by man (such as grain) is said to be much less: around 0.05% according to Southwood (1972). Further losses occur when plant matter is consumed by a herbivore: the conversion rate here is put at a maximal 10–25% according to species and stage in the life cycle. This phenomenon is accentuated by the fact that energy flowing to higher trophic levels is raised to higher free energy levels (animal bodies contain more fat, a concentrated energy form, than plants) and a good deal of the energy in the lower trophic levels is degraded to raise the rest of the material (Morowitz, 1968). From this stem two main points: first that there would appear to be room to increase the production of food for man by enhancing the efficiencies of these energy conversion processes (and much has been done) and second that in an ecosystem the biomass (that is, living weight) of animal matter is always less than that of plant material, and that of carnivores less than that of herbivores (Figures 2 and 3). Similarly, the

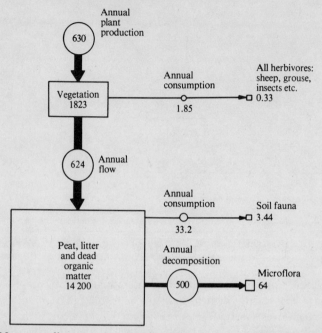

Figure 3. Mean standing crop (rectangular boxes) and production/consumption (circles) in a moorland ecosystem at Moor House, Cumbria. All figures are grams per square metre. From Heal and Perkins (1975).

production of plant foods for man requires less space than the production of equivalent weight of meat protein (Table 2: Yapp and Watson, 1958; Ovington, 1963).

1.2.3 Succession

Not only do the proportions of the major components of ecosystems, and the apportionment of energy between them, vary from system to system but there are often well marked changes with time. Sometimes these are in response to changing climate, as when receding ice some 20000 years ago was followed by the northward movement in Europe and North America of tundra, birch and pine forest and broad-leaved deciduous forest with their attendant animals. Other changes may be caused by the components of the ecosystems themselves. This is the basis of the classic concept of succession first developed by F. E. Clements (1916), in which dominant plants are envisaged as altering the system to make it more favourable to other species and less suited to themselves. Today this kind of 'relay floristics' is considered to be only one of several patterns (e.g. Connell and Slayter, 1976; SCOPE, 1977). In others, dominant plants may

Table 2. *Area of land needed to produce 50 kg of protein in one year under various crops*

Plant crops	(hectares)	Animal husbandry	(hectares)
Beans	0.1	Dairy cows	0.45–1.1
Cereals	0.23	Sheep	0.8–1.9
Grass	0.1–0.25	Beef cattle	1.0–25
Potatoes	0.27	Fowls	1.44
		Pigs	2.2

After Bender (1963).

so modify the system as to exclude other colonists (Figure 4). In either case, human actions, either directly through removing the dominant species or subjecting it to a stress such as grazing, or indirectly through altering water regime or chemical factors, can set in train rapid changes. Such impacts may be intensified by the fact that it is probably rare for there to be just one dominant species of plant capable of success in a particular habitat. When an area is opened up there may be something of a race between potential colonists, and the character of the ecosystem that develops may depend on which succeeds in establishing itself first. A disturbed system may thus not be recolonised by the original dominants (Connell and Slatyer, 1976).

1.2.4 Life strategies

Very broadly, there are two kinds of life strategy in plants and animals (with all manner of intermediates), and these are sometimes termed ' r ' and ' K ' strategies, following terms from the logistic equation of population growth (Southwood, 1976). The first type is characterised by a high reproductive rate: many offspring are produced but there is little care of the young, the adults are not long-lived, and are generally small in size, and the increment of growth in a year is large in relation to the 'standing crop' or biomass of the population. The strategy is well equipped to provide rapid multiplication in a new habitat or favourable period, but is not suited to withstand a long period of adverse conditions or intense competition. 'Boom and bust' cycles are typical of such organisms, with population peaks and crashes like those familiar in many insects or in rodents like lemmings. The species are continual colonists and often migrants. They are familiar as weeds and pests in ecosystems disturbed by man.

In contrast, K-strategists are generally longer lived, larger, with lower

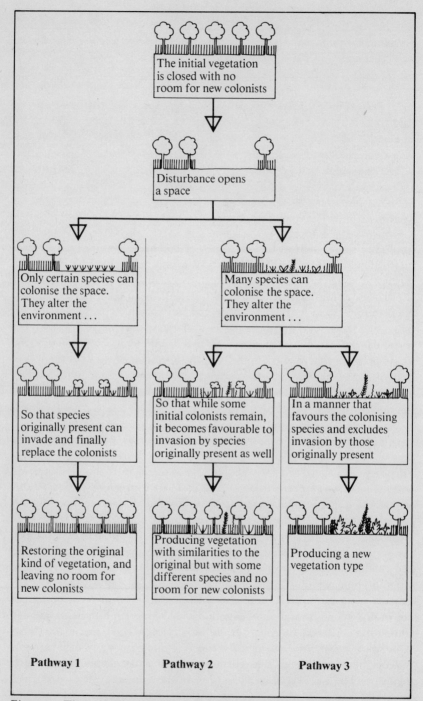

Figure 4. Three kinds of succession restoring vegetation to a disturbed area. Pathway 1, the traditional successional system, is termed the 'facilitation' model, pathway 2 the 'tolerance' model and pathway 3 the 'inhibition' model. Modified from Connell and Slatyer (1976).

reproductive rates and ways of caring for the young. They are typical of stable habitats and have evolved so as to maintain their populations at an equilibrium level in the face of intense competition. They are adapted to withstand normal environmental fluctuations, especially since they do not have to go through a reproductive cycle every year. A larger biomass is contributed by a smaller number of individuals, and energy is used more efficiently. Many of the trees, and all the large mammals that dominate long-established ecosystems are K-strategists.

These strategies confer differing degrees of resilience on their followers and make them respond differently to pollution and other stress imposed by man. Dominant K-strategists resist stress within certain limits, but if perturbed beyond these may be unable to recover quickly, thus creating openings for new invaders. These will often be opportunist r-strategists, producing plant and animal communities that are very different to those originally present.

1.2.5 *Stress and resilience*

There has been much debate in recent years over the relationship between the diversity of an ecosystem (diversity being defined in terms of the number of species present) and its stability. It has been suggested (e.g. Elton, 1958) that diverse systems will be less likely to fluctuate widely in response to climatic change or disease because so many species are present that if the population of one is temporarily depressed there are likely to be others filling comparable, though not identical, ecological roles that can multiply and sustain the character of the whole. In contrast, in a species-poor system the loss of a single dominant may cause a much greater change. However, this is likely to be an over-simplification. In relatively constant and favourable environments, the ecosystems that develop following what are often lengthy successional processes generally contain a large number of K-strategist organisms adapted to withstand the normal range of fluctuations, and recover swiftly from temporary disturbances. The system thus has considerable resilience. But stressed beyond those limits, disruption can be very great, at least turning the system back towards the kind of configuration present in early successional stages (Grime, 1973; Regier and Cowell, 1972) and often substituting quite different systems. On the other hand, systems like those of deserts or polar regions may be subject to continual physical stress, and the dominant species are more likely to be r-strategists, capable of rapid exploitation of brief favourable periods: the number of species is generally much less although those that do occur may be highly adapted and consistently present. Such ecosystems have a different kind of resilience and apparent 'stability', yet may be equally vulnerable to the added stress of human disturbance.

Whatever the system, a major aim of environmental management and pollution control is to avoid stressing a valued ecosystem – whether wilderness, farmland or forest – beyond the limits of resilience of those species whose perpetuation is desired. The standards and objectives discussed in Chapter 6, and the controls decribed in Chapter 8, need to be based on an understanding of the ecology of potential 'target' species.

1.3 *The impact of man*

1.3.1 *The stress of pre-industrial man*

Early man developed within, and ranged over, many ecosystems and by the beginning of recorded history was established in all the main biomes of the world. In the earliest period, when our ancestors were 'hunter-gatherers' living on those edible plants they could find in the wild and the more readily collected animals foods (like eggs, shellfish or large terrestrial invertebrates) their numbers were probably low and their impact small. None the less, three kinds of environmental stress can be recognised even among people who live in this simple fashion.

The first, which may be termed 'eutrophic stress' is caused by the deposition of human body wastes and food residues in small areas where concentrations of nutrients can rise sufficiently to affect plant growth and the balance of ecosystems. Phosphate and nitrate levels are sufficiently raised around some sites of early settlement to make them useful as an indicator to archaeologists today (Svensson and Soderlund, 1976; SCOPE, 1977). Eutrophication (the word means 'enrichment') of fresh waters by the much more voluminous body wastes of urban man remains a major modern pollution problem and is described as such in Chapter 5.

The second impact may be termed 'exploitative stress'. It arises directly from the cropping of wild animal and plant populations (Le Cren and Holdgate, 1962). Any organism in nature has the capacity to produce more offspring than are needed to replace the parents, and even in K-strategists there is normally a considerable juvenile mortality. Predation by man, like predation by other carnivores, does not threaten the survival of the prey species unless it exceeds its capacity to make the loss good (the 'sustainable yield'). Short of this point, cropping alters the structure of the exploited population, whose members tend to commence breeding earlier and live less long (both, as would be expected, movements towards an r-strategy). Cropping beyond the sustainable yield, if protracted, equally naturally leads to a collapse in the prey population. Early man may well have exterminated various very large land animals with low reproductive potential (extreme K-strategists) like the giant ground sloth of South America (*Mylodon*) or the mammoth (Martin and Wright, 1967). Later years have seen the Dodo and most recently, certain species of great whale, brought low in the same way (although not, in the latter, to actual extinction).

The third basic impact may be termed 'disruptive stress'. It involves the modification of the structure of an ecosystem – for example by fire – so that areas are opened up to invasion by fresh individuals (often of *r*-strategist species not prominent in the previous situation) and new successions which may or may not restore the original ecosystem are begun. Early man probably used fire to help drive potential prey animals, quite apart from accidentally burning vegetation when cooking fires got out of control. The role of disruptive stress increased as hunter–gatherer societies were succeeded by pastoral and agricultural ones.

1.3.2 *The development of agriculture*

Traditional shifting or 'slash and burn' cultivation in forest zones is, perhaps surprisingly, a relatively successful way of cultivating herbaceous food plants without over-stressing the forest ecosystem as a whole. In it, trees and shrubs are cut at the end of the dry season, and after 10 days or so they are burned. This releases nutrients, especially potassium, calcium, magnesium and phosphorus, to the soil. The enriched soil is then cropped intensively for up to 3 or 4 years. In that period there is a rapid oxidation of up to 5 tonnes of humus and old roots per hectare per month. At the end of the period yields on the depleted soil begin to fall, weed (*r*-strategist) control problems increase, and the plot is abandoned to be regenerated slowly, maybe over a century. Given plenty of land, so that the cultivation patches are always surrounded by plentiful forest seed-parents, and plenty of time can be allowed for the succession back to forest, the process need not be inefficient or lead to long-term ecological change (Conway and Romm, 1972).

Larger-scale extensive use of fire to clear forest is another matter. Such clearing by burning has dramatic ecological effects. It destroys at a stroke a large part of the organic matter that was locked up in the 'standing crop' of the forest, passing it back into the air as carbon dioxide, oxides of nitrogen and water vapour. It releases a large quantity of nutrients. It may also burn much of the leaf litter of the forest floor and the upper organic layers of the soil, with a further release of nutrients, and it opens these soils to wind and rain at a time when, devoid of their vegetation cover, they are especially prone to erosion. The result can be a significant nutrient loss that may require many centuries of soil maturation of a major input of fertiliser to put back. One experiment (Likens, Bormann, Johnson, Fisher and Pierce, 1970) demonstrated a net loss of 65–90 tonnes of eleven nutrients/(km² year) in cut forest as against 4.5–6 tonnes from uncut forest. The latter levels could, but the former could not, be made good by the weathering of the rocks and the deposition of dissolved salts in rainfall. If domestic stock grazing is allowed to develop too fast, the grassland that succeeds the forest may never completely bind the soil, and erosion may persist.

Pastoralists convert forest into herbaceous, especially grassy vegetation. The former is dominated by large, long-lived trees in which the net amount of new living matter produced each year is small in relation to the standing crop, and in which much of the nutrients in the system are locked up in the standing crop. In the vegetation that is substituted, the annual production is large in relation to the standing crop and nutrients are cycled more swiftly. The pastoralists also replace large, wild *K*-strategist herbivores, like bison, and predators like wolves that took their surplus by more manageable stock like cattle and sheep, with man as the sole predator. Agriculturalists extend this process by growing selected plants, which usually have annual cycles and need to be *r*-strategists since we generally eat their seeds or fruits. With both crops and livestock the principle of sustained yield is important. If the cropping rate exceeds the capacity of the soil to replenish essential nutrients, and man does not return nutrients – anciently by fertilising the ground with his own body wastes, the wastes of his livestock and the residues of the crop that he does not eat – fertility will decline. Most natural and undisturbed vegetation has a low and sometimes intermittent nutrient flow and overall balance: in woodland some 90% of nutrients are cycled within the system. In agriculture, nutrients are taken away, especially in modern farming whose produce is removed to the cities from where the nutrients flow through sewers to rivers or the sea.

If ecological systems are to be modified extensively for food production, management effort and especially the deliberate application of fertilisers to balance nutrient losses in the crop must therefore increase. This increasing application of fertilisers to land on which a single crop species is grown over large areas has been a dominant feature of agricultural history. The production of useful food per unit area has increased dramatically in consequence, as Table 3 shows. But as the intensity of management has risen, so agriculture has become more dependent on urban technology, driven by fossil fuel energy, to supply the fertilisers and equipment that make the intensive cultivation possible (Bourne, 1973).

Such crops also need artificial protection by man. A crop monoculture is species-poor by design. Open ground is deliberately maintained by disturbance. This ground is readily colonised by quick-growing, fast-multiplying *r*-strategist 'weeds'. The crop itself offers great opportunities to *r*-strategist herbivorous insects or parasites specialising in the species grown, because there is so much of it in one place and because the habitat changes tend to remove from the system the predators that might limit

Table 3. *Wheat yields in Britain since 1200*

Date	Event	Yield (tonnes/hectare)
1200	Open field system	0.5
1650	Enclosure, fallowing	0.62
1750	New seed drills	1.0
1850	Four-course rotation	1.76
1900	Fertilisers. New varieties	2.13
1948	More fertilisers. Selective weed killers	2.51
1958	Short straw varieties	3.13
1962	Increased use of nitrogenous fertiliser	4.27
1977	New varieties and greatly improved techniques	up to 10.0

After G. W. Cooke (1970).

the populations of the potential 'pests'. (The complexities of the 'pest' situation have been analysed most illuminatingly by G. R. Conway, 1976.) Hence the need for more management, and for pesticides. Some of these toxic substances can have side-effects on wildlife, further reducing the natural enemies of the pest. The result is a reduction in the natural checks and balances and increased dependence on management, supported by technology and by supplementary energy.

1.3.3 *Industrial stress*

Eutrophic, exploitative and disruptive stress, accompanying man's social evolution and agricultural development, are quite evidently counter-productive if carried too far, for they undermine the capacity of wild or modified ecosystems to produce food and sustain people. Too much enrichment of the environment with sewage can destroy fisheries as well as making lakes, rivers or estuaries smelly and unhealthy. Over-cropping brings its own swift reward in the failure of the resource thus misused. Excessive disruptive stress can waste fertility and demand needlessly costly human corrective effort. The groups of people imposing such stresses thus have a direct interest in regulating them, short of the point of severe ecological damage.

But the fourth kind of stress which has been imposed on ecosystems as a consequence of the urbanisation and technological development of human societies – here termed 'chemical and industrial stress' – is different in kind. It arises from activities like energy generation, metal smelting and the multifarious aspects of manufacturing industry. These activities bring undoubted social benefits: modern developed countries are founded upon industry. But these industrial processes do not depend for their benefits upon preserving the productivity of ecosystems. Indeed, the replacement of biologically productive soil by unproductive substrata of concrete or stone is a normal feature of industrial expansion into 'green field' sites. There is thus no direct feedback working to restrain the impact of the stress. Such feedback comes in part because of the effects of this stress on people working within the industries concerned, and partly because the substances that are released into the environment may affect land at a distance, whose biological productivity remains important. In the remainder of this book, an attempt is made to describe the kinds of pollutant involved both in eutrophic and in chemical and industrial stress, and to evaluate the scale of their impact directly and indirectly on living targets and biogeochemical cycles.

1.4 *The context of pollution*

While most of the rest of this book is about pollution and our social response to it, the context must not be lost. Pollution is only one of the ways in which man alters the natural world. It operates through changing the levels – and sometimes the natures – of the chemical and physical factors on whose balance the composition and productivity of living systems depends. In some respects its effects on those systems parallels those of other types of human disturbance, like fire, overgrazing and clearance for agriculture. This is because all impose stress on ecosystems and change their balance, generally promoting the replacement of large, long-lived K-strategist plants and animals, and of communities with a high level of diversity and competition, by others with a lower diversity, lower competitive level, and higher proportion of r-strategists (Regier and Cowell, 1972; Grime, 1973; see also Chapter 5). In considering the impact of pollutants these wider ecological interactions should not be forgotten. For our reactions to pollution depend on the value we place upon the systems it alters, and our policies to prevent or control it should always be seen as but one component in the wise stewardship of the natural resources of the earth.

2. Pollution and pollutants

2.1 *Definitions*

Pollution may be defined as:

The introduction by man into the environment of substances or energy liable to cause hazards to human health, harm to living resources and ecological systems, damage to structures or amenity, or interference with legitimate uses of the environment.

This definition is based on a more restricted one evolved by the Group of Experts on Scientific Aspects of Marine Pollution (GESAMP, 1972). It also closely follows the definition used in Article II of the Paris Convention on the Prevention of Marine Pollution from Land-Based Sources, and has something in common with the terminology used in documents of the United Nations Conference on the Human Environment (the Stockholm Conference) and the United Nations Environment Programme (UNEP).

The important concepts in this definition are:

(a) Pollution is caused by substances or energy;
(b) It has a source or sources, and they are created by man. Natural inputs of the same substances are excluded. Thus pollution is an increment added by man to biogeochemical cycles;
(c) Pollution acts in the environment, as a result of these discharges, and follows a pathway, leading to the exposure of structures or organisms;
(d) The significance of the pollution is related to its effects on a range of targets (or receptors), including man and the resources and ecological systems on which he depends;
(e) Pollution is judged by its impact on social values as well as environmental components: if there is damage to structures or amenity, or interference with legitimate uses of the environment, the substances causing the effect are by definition, pollutants. (For this reason, bulky debris liable to foul fishing gear comes within the scope of the 'black list' of materials that may not be dumped at sea under the Oslo Convention);
(f) Quantification of the scale of the hazard or damage or interference is important, and the basic question is one of acceptability of the consequences of the release of the substances or energy.

17

In the simplest possible language, pollution is 'something in the wrong place at the wrong time in the wrong quantity'. This more superficial generalisation still enshrines the basic essentials: that pollution is caused by a substance or energy, that the location of the pollutant and its effects in space and time are crucial, that the quantity of pollutant and scale of effect must be evaluated, and that a value judgement about what is acceptable (or 'wrong') is inevitable. Pollution has to be seen in a socio-economic context. That is why there is so much argument about it.

2.2 *Pollution by energy and sound*

Energy is discharged into the environment in several potentially damaging forms – most commonly as heat and noise, but also as nuclear, ultra-violet and other types of radiation and as inaudible low-frequency sound. Such physical pollutants are generally classified separately from the chemical pollutants, which exceed them greatly in diversity of nature and effect and receive most attention in this book.

'Nuclear' or 'atomic' radiation is ionising: that is it is able to alter the structure of atoms within targets receiving it, with considerable potential biological effects. There is a significant natural background level of such radiation, from rocks that contain radioactive minerals and from the sun and other 'cosmic' sources. Living organisms do not discriminate between radioactive and non-radioactive materials when building up their tissues and hence contain very small quantities of naturally-occurring radio-isotopes. As a consequence, if the concentrations of these substances rise in the environment, more will also find their way into living tissues.

Human exposure to ionising radiation has increased in several ways over the past century. One of the first fairly widespread, although generally small, increases came from the introduction of X-rays into medicine, before the need for the present tight controls was recognised. Between 1945 and about 1965 nuclear weapon explosions in the atmosphere added significant quantities of radioactive substances to the environment. More recently, low-level emissions from nuclear power stations and the associated fuel processing works have become proportionately more important, and this trend seems likely to continue (Saunders, 1976; Royal Commission, 1976b). The disposal of wastes from these power stations, and of radioactive materials from hospitals, laboratories and other sources, is very strictly controlled so as to minimise environmental contamination, and there is a considerable debate over levels of acceptability which need not be pursued here. What is evident is that as a result of these measures, the amount of ionising radiation the average person receives as a result of human activities is unlikely to rise significantly above present levels, which are no more than a few per cent of the natural background. However, because of the nature of the effects of such radiation (see Chapter 5), even small increments are considered undesirable in principle and stringent controls to reduce them to a minimum will undoubtedly be maintained.

The other types of pollution by energy are more commonly experienced but cause less concern. The laws of thermodynamics make it inevitable that much of the energy we generate ends up as waste heat. Man thus warms his environment when he generates power. Chapter 4 describes the effects of this heat discharged to the air and the aquatic environment, indicating that the changes caused are local and at present of no great magnitude. None the less if power generation continues to grow at the rate suggested by projections like that in Figure 17, the environmental effects could become more significant (see Section 4.3.1).

Noise is the most widely experienced, and probably the most annoying, of all physical pollutants. It is rarely loud enough to cause measurable biological damage to man, except in some working environments where protective measures are not properly observed, and in some places of so-called entertainment. But in the neighbourhood of some airports, factories and busy highways it may reach levels capable of causing irritation and stress, and has accordingly provoked control measures, discussed briefly in later chapters.

2.3 Pollutant substances

As Chapter 1 indicated, living systems are in a state of continuing interaction and change. Stability, far from being the expected manifestation of some ideal 'balance of nature', is an anomaly in most environmental systems, and where equilibria persist for long periods, they are the outcome of many interacting and dynamic processes. Organisms and ecosystems are continually adjusting to external, interacting, physical, chemical and biological factors.

These factors include a vast number of chemical variables, including the availability of carbon, oxygen, nitrogen, phosphorus, sulphur, magnesium, calcium, manganese, boron, iron, cobalt, nickel, copper, zinc, potassium, sodium and many other elements, sometimes alone (e.g. oxygen) but usually in a range of compounds.

Many of these elements are essential and non-toxic, and hence non-polluting in certain formulations, but hazardous in others. Carbon and nitrogen, for example, are essential to plants as carbon dioxide, nitrate or ammonium but toxic to animal respiratory systems when combined as cyanide and presented at relatively low concentrations. Plants require copper, iron, manganese, zinc, boron, molybdenum and cobalt – but excessive quantities of copper or zinc can sterilise soils around old metal workings or smelters. Other substances such as cadmium, mercury and lead, are thought to be non-essential to both plants and animals but occur naturally in the earth's crust and are not necessarily toxic – indeed they are normally present, at low concentrations, in living organisms.

Toxicity depends on formulation and concentration. Just as the combination of carbon and nitrogen into cyanide confers toxicity on normally harmless elements, so the form in which mercury or lead exists in the environment is all-important. The organic combinations of mercury – especially methyl mercury – pose a greater hazard than its inorganic salts. Similarly, sulphur is an essential component of all living organisms: sulphur dioxide is a naturally occurring gas in the atmosphere, yet high concentrations in urban zones can undoubtedly cause damage to plants and people. It is very important, when evaluating the possible significance of environmental contamination with a substance, to determine the form in which it is present, and the amount present. It is also important to recognise that different substances pose a hazard to different targets: as a result there is commonly a possibility of differential effects within a complicated ecosystem, causing changes in the relative abundance of the component plant and animal species.

One distinction that is sometimes drawn is that between 'natural' and 'synthetic' substances. The former include inorganic materials like lead, mercury or sulphur dioxide, and organic substances of biological origin like urea or oil. All have been present in the environment throughout most of evolutionary history. Some – and this still applies to many well-known pollutants like sulphur dioxide, carbon dioxide and carbon monoxide – are still produced by natural processes in quantities as large or larger than are released by man. Some are the inevitable consequences of organic life, like faeces, urine and the products of bodily decay. Such basic biological pollutants are inevitably released by man in mounting quantities as his population rises and as the number of livestock he keeps increases. Associated basic non-biological pollutants like the products of fuel combustion are an almost equally inevitable consequence of even simple human life. On the other hand, man has greatly increased the output to the environment of many such naturally occurring substances because they are also generated in industry – including, again, the products of fuel combustion and metals and inorganic salts extracted and processed for a variety of uses.

At the other end of the scale come the truly synthetic materials which man has produced for the first time – materials like organochlorine pesticides, many pharmaceutical and cosmetic products, and plastics. Some of these (like DDT) are useful because they are not only biologically active but stable, persisting for a long time in the environment. Some ecologists are especially concerned about the impact of these materials because organisms have not been exposed to them during the hundreds of millions of years of evolutionary history. Their long-term effects are something of an unknown quantity. Toxicity testing in the laboratory can and does provide a screen against dramatic, acute harm to man or livestock, but there remains a suspicion that such substances might have unforseen effects on living creatures over longer periods of exposure than we can possibly test for, or upset the balance of ecological systems far from the places of usage.

Pollutants are defined in this book as substances causing damage to targets in the environment. This definition excludes potentially hazardous materials used by people on themselves – like cosmetics, food additives, pharmaceuticals or, notoriously, tobacco smoke. Many of these materials are synthetic, and new products are being evaluated every year. The tests to which they are subjected are becoming increasingly rigorous, partly because improved methods are continually devised and partly because experience (for example with thalidomide) has demonstrated the inadequacies of former procedures. These tests have traditionally (and understandably) been related to the main circumstances of use rather than the possible behaviour of the products when discharged to the environment, but there is an inevitable overlap in the knowledge required to evaluate safety under both circumstances and both are now commonly considered in the screening process described later in this chapter.

This exclusion of 'materials used by people on themselves' has a consequence of great importance. Recent evidence suggests that a high proportion of cancers – perhaps as many as 60–80% – may be provoked by 'environmental' factors rather than result from genetic defects and biochemical instabilities. Similarly, many cases of teratogenesis (the production of abnormal embryos) can be linked to exposure of the mother to organic compounds like dioxins (as contaminants in herbicides and other products), hexachlorophene (used as a disinfectant) and organic solvents. But the 'environment' for carcinogenesis is the whole outside world beyond the body, and the statistic quoted above does include the effects of smoking, food contamination, dietary indiscretion (a polite clinical description of excessive drinking and the like), industrial and occupational exposure, drugs, cosmetics and disinfectants. Probably these predominate in the overall pattern. Certainly recent trends in cancer provide no evidence of an upsurge that might be related to the increase in the amount and diversity of chemicals emitted to the environment over the past 50 years. There may be a 'pollution effect' in the generally higher incidence of some cancers in urban areas, as there may be in the relationship between the softness of drinking water and the incidence of cardiovascular disease, but recent advances in medical understanding imply that environmental pollutants, strictly defined, are likely to be less important than substances in the more immediate human environment of work, food and self-indulgence as causes of illness. This is a theme returned to in Chapter 10, and documented in the report of a recent Royal Society Discussion (Doll and others, 1978).

2.4 Classification of pollutants

Almost any biologically active substance – and there are few substances which have no effect on the biochemistry of living organisms at some concentration – can behave as a pollutant. It is not surprising that many alternative ways of classifying this diversity of materials have been employed. They include those set out in Table 4. Several general points are evident from this long, but far from comprehensive list. First, the classifications are not mutually exclusive. Indeed it is common to adopt a matrix type of classification. Very often a physical state (e.g. gas, liquid or solid) is used as a primary unit to group chemical substances. Alternatively, environmental sectors (air, fresh water, sea, soil) may be chosen, and subclassifications based on source, substances or effects. A third variant uses targets as major units, considering substances hazardous to man, livestock, farm crops or particularly vulnerable or important species and ecosystems. As a fourth approach a class of substance, such as detergents, may be selected and a subclassification based on properties like biodegradability. Second, there is no 'ideal' classification. Classification is for a purpose, and the selection must be use-related. Each system is convenient within certain limits. The value of the classification is as an aid in the orderly assembly of information. It also helps to indicate gaps in knowledge, because certain units are recognised as logically necessary but when the available information is apportioned, some 'pigeon holes' remain empty. This is the rationale behind recent tabulations of pollution research by the Natural Environment Research Council (NERC, 1976), the Inter-Research Councils Committee on Pollution Research (IRC-COPR, 1971–77) and the Department of the Environment (DOE, 1975a).

Almost all classifications in this field are arbitrary and to a degree subjective. The most 'natural' classifications are those based on the chemical and physical properties of substances: moreover these readily relate to the general body of chemical knowledge, and can accommodate the distinction between 'natural' and 'synthetic' substances mentioned above (which, however, is not readily developed as a classification in its own right). Classifications by effect are 'natural' in that they relate to the natural classification of living things and can also be linked to the general body of biological and especially physiological knowledge. The classification based on steps in the pathway linking source and effect is more artificial, for substances move freely between sectors and few types of human activity have a monopoly as sources of particular materials (see e.g.

Table 4. *Alternative systems for classifying pollutant substances*

(A) Classification by nature

(*a*) *Chemical composition*
  (i) Inorganic pollutants (listed in order by elements and compounds e.g.)
  carbon particles (smoke)
    carbon monoxide
    carbon dioxide
  sulphur
    sulphur dioxide
    sulphuric acid
    sulphate (ion)
  etc. (the list is a very long one)
  (ii) Organic pollutants (listed according to standard chemical groupings, e.g.
  hydrocarbons, alcohols, esters, ketones etc.) (this is an even longer list)

(*b*) *Physical state*
  (i) Gaseous pollutants
  (ii) Liquid pollutants
    as natural liquids
    as solutions in other substances
  (iii) Solid pollutants
    as solid wastes (various types, including complex mixtures)
    as suspensions of particles in air or water or some other medium
      (particle size subclassification)

(B) Classification by properties e.g.
  solubility in water, oil, fat
  rates of dispersion and dilution
  biodegradability
  persistence in air, water, soil or living organisms
  reactivity with other substances

(C) Classification by sectors of environment
  air pollutants
  fresh-water pollutants
  marine pollutants
  soil or land pollutants

(D) Classification by source

(*a*) *Products of fuel combustion*
  domestic sources
  industrial sources
  agricultural and forest sources
  vehicular sources

(*b*) *Products of industrial origin*
  (classify by industrial process e.g. cement industry, ceramics
   industry, sinter plants)
  (commonly discriminate between or subclassify into processes
   emitting to air, rivers, sewers or sea or generating solid residues)

Table 4 (*cont.*)

(*c*) *Products of domestic or institutional origin*
  human body wastes
  kitchen wastes
  household solid wastes
  hospital wastes
  laboratory wastes
  trade wastes

(*d*) *Products of agricultural origin*
  residues of intensive livestock husbandry
  run-off from fertiliser application
  dissemination of pesticide residues

(*e*) *Products of military activities*

(*f*) *Products of microbiological or fungal activity*
  aflatoxins and mycotoxins

(E) Classification by patterns of use e.g.

(*a*) *Uses in industry*
  raw materials
  construction materials (including protective materials such as paints)
  solvents
  catalysts
  plasticisers
  stabilisers
  preservatives
  coolants
  lubricants
  detergents
  transmission fluids
  insulation materials (fire and electrical insulation)
  pesticides etc. (the list may need adjustment from industry to industry)

(*b*) *Uses in the home or in hospitals, schools, hotels etc.*
  solvents
  coolants (refrigerants)
  food additives and preservatives (including colouring matter)
  pesticides
  detergents
  insulation materials
  construction materials (including paint etc.) etc. (there is an
    inevitable overlap with the industrial list)

(*c*) *Uses in agriculture and horticulture, and by local authorities e.g.*
  in pavement or swimming pool maintenance
  fertilisers
  pesticides
  sterilants etc. (within buildings and in connection with equipment there
  is inevitably overlap with the two preceding)

Table 4 (*cont.*)

(*d*) *Uses in transport*
   fuels
   lubricants
   vehicle construction materials including transmission fluids
   maintenance materials (including cleaning materials, anti-fouling and
      other paints etc.)
(*e*) *Uses in defence*

(F) Classification by target and effects
   substances affecting atmospheric processes e.g. ozone layers,
      radiation transmission
   substances affecting processes in water
   corrosive agents
   substances affecting man directly (subclassify by nature of effects, e.g. substances
      blocking various enzyme systems, mutagens, substances producing abnormal
      embryos (teratogens), carcinogens, substances affecting central nervous system,
      substances affecting endocrine system and hormone balance, substances
      impairing respiration, substances affecting circulation, substances affecting
      kidney and excretion, substances affecting genetics and reproduction)
   substances affecting livestock (similar classification by physiological action)
   substances affecting non-domestic animals (similar subclassifications)
   substances affecting crop plants and commercial trees (subclassify by nature of
      physiological effect, e.g. substances damaging foliage, impairing photosyn-
      thesis, damaging root system, impairing nutrient or water uptake or trans-
      mission, substances affecting growth and form, substances affecting genetics
      and reproduction)
   substances affecting wild plants (similar subclassification)
   substances affecting ecological systems (subclassify by nature of effects)

Table 5): the classificatory units relate to the industrial and administrative structure of society rather than to any particular element in the natural world. Which system is chosen thus evidently depends on the purpose for which the information is being brought together.

2.5 *Pollutants in environmental sectors*

2.5.1 *Pollutants in the air*

One of the commonest general classifications is by sector, and by source within that sector. Figure 5 arranges the common pollutants of air in a matrix of source against physical state. Many of the substances commonly regarded as particularly troublesome are products of 'fossil' fuel combustion – carbon dioxide, carbon monoxide, unburned hydrocarbons, oxides of nitrogen, sulphur dioxide, and smoke, together with the products of

Table 5. *Some emissions from particular industries*

Steel mills: particles, smoke, carbon monoxide, fluorides
Non-ferrous smelters: sulphur oxides, particles, various metals
Petroleum refineries: sulphur compounds, hydrocarbons, smoke, particles, odours
Portland cement plants: particles, sulphur compounds
Sulphuric acid plants: sulphur dioxide, sulphuric acid mist, sulphur trioxide
Iron and Steel foundries: particles, smoke, odours
Ferro-alloy plants: particles
Pulp mills: sulphur compounds, particles, odours
Hydrochloric acid plants: hydrochloric acid mist and gas
Nitric acid plants: nitrogen oxides
Bulk storage of petroleum produces: hydrocarbons
Soap and detergent plants: particles, odours
Caustic and chlorine plants: chlorine
Calcium carbide manufacturing: particles
Phosphate fertiliser plants: fluorides, particles, ammonia
Lime pl nts: particles
Aluminium ore reduction plants: fluorides, particles
Phosphoric acid plants: acid mist, fluorides
Coal cleaning plants: particles

After American Chemical Society (1969).
This was based in turn upon hearings before the Sub-Committee on Air and Water Pollution of the Committee of Public Works, US Senate, 90th Congress, 1967.
Note: The order in this list does not necessarily correspond to relative quantities emitted or to the seriousness of the effects.

Table 6. *National air pollutant emissions, millions of tons per year, 1965*

	Totals	% of totals	Carbon monoxide	Sulphur oxides	Hydro-carbons	Nitrogen oxides	Particles
Automobiles	86	60	66	1	12	6	1
Industry	23	17	2	9	4	2	6
Electric power plants	20	14	1	12	1	3	3
Space heating	8	6	2	3	1	1	1
Refuse disposal	5	3	1	1	1	1	1
Totals	142		72	26	19	13	12

From 'Cleaning our Environment: the Chemical basis for Action', based in turn on *The Sources of Air Pollution and Their Control*, Government Printing Office, Washington, DC (1966).

chemical reactions between them, such as peroxyacetyl nitrates (PAN) and other oxidants. This group of substances has been the subject of nationwide inventories in several countries (e.g. US Department of Health, Education and Welfare, 1968; American Chemical Society, 1969;

Medium	Physical state	Substances and sources			
		(1) From fuel combustion (including vehicles)		(2) From industrial processes	(3) From natural sources (volcanoes, decomposition)
		Primary	Secondary derivatives		
Air	Gaseous	Carbon dioxide Carbon monoxide Unburned hydrocarbons Oxides of nitrogen → [PAN Ozone and other oxidants] Sulphur dioxide → [Acidity in rainfall] Aldehydes (main causes of smell of car exhaust)		Organic vapours, chlorofluoro compounds Acids, acidic aerosols Aldehydes (common cause of smells) Fluorides, chlorides, bromides etc.	Carbon dioxide Carbon monoxide Ozone Methane Hydrogen sulphide (oxidised to sulphur dioxide) Sulphur dioxide Oxides of nitrogen
	Particulate	Smoke Grit and dust Lead particles (aggregated with carbon from vehicle emissions)		Grit, dust, metallic particles, metallic oxides, carbon particles	Dust

Figure 5. Common atmospheric pollutants.

IRCCOPR, 1974) and Table 6 records the estimated emissions from various sources in the United States in 1965 (see also Kneese, Rolfe and Harned, 1971 and CEQ reports). Excess heat is also produced (see Chapter 4): water, also a universal combustion product is neither generally mentioned nor problematical. As Figure 5 shows, many of these compounds are also produced naturally by volcanoes and biological decomposition. Their widely differing effects, and the trends in their global and regional concentrations, are considered in Chapters 4 and 5.

Industrial processes generate a much longer list of substances, many of which are vented to the air. Only a few are listed in the middle column of Figure 5. The vast range of hypothetical problems can be illustrated by examining some of the lists of actual or potential air pollutants compiled by various expert groups. Two hundred and sixty-one 'potential atmospheric contaminants' are cited, for example, by Sheehy, Achinger and Simon (1969). The absence of a substance from these lists does not mean it poses no problem: it may be a new product since the list was prepared or no environmental effect may have been detected (possibly because none has been looked for). What such lists do more than anything is demonstrate that catalogues of the names of substances are virtually impossible to interpret unless they are amplified by information about the properties and effects of each component.

2.5.2 *Pollutants of water*

There are just as many common pollutants of fresh waters and the sea. Some are set out in Figure 6, and the general field has been reviewed by several authors (e.g. Mackenthum, 1969). About 1500 commonly encoun-

Medium	Physical state	Substances and Sources			
		Domestic			
		Primary	Secondary	Industrial	Agricultural
Fresh waters	Dissolved	Organic body wastes $\longrightarrow$ Detergents $\longrightarrow$ Pharmaceuticals Cosmetics Pesticides Metal salts	Nitrates Phosphates Carbonates Phosphates	Very wide range of organic substances: detergents pharmaceuticals oils pesticides metal salts	Concentrated organic livestock wastes: phosphates nitrates Pesticides
	Suspended	Organic and inorganic particles		Organic and inorganic particles	Organic and soil particles

Figure 6. Common pollutants of fresh waters.

tered substances are listed in the Index of Solubility, Toxicity and Biodegradability (INSTAB), maintained by the British Water Research Centre, and continually up-dated. They have many origins. Historically, the first to have a significant impact on the environment and to pose a hazard to human health and to fisheries may well have been the human body wastes whose role as producers of eutrophic stress was mentioned in Chapter 1 and is further described in Chapter 5. However, some toxic materials also arise from ancient human occupations. Mining stands high on the list. It releases sediments which, borne upon swift-flowing waters, are deposited when the current slows on entry to a lake, lowland river or estuary. This sediment may blanket the life on the bed of the water body and so kill it, or it may alter flow patterns. Mining also produces acid wastes, and some of these are rich in metals and other elements like lead, copper, zinc, arsenic, mercury or iron. Many of these are toxic at fairly low concentrations, and the impoverishment of the fresh-water life of rivers in areas traditionally mined for metals, for example in Wales and Derbyshire, is well known (Jones, 1940; Pentelow and Butcher, 1938; Hynes, 1959; Glover, 1975). Living organisms – especially molluscs, but also Crustacea and some plants like the water buttercup, *Ranunculus fluitans* – have the capacity to concentrate metals many hundredfold and hence to raise levels which may be of no significance in the water to the point at which their tissues become highly hazardous to things that eat them. Lead, cadmium, mercury and arsenic are particularly troublesome because of this.

Other toxic substances are the products of modern industry, and many are new synthetic materials. Detergents, pharmaceutical products and pesticides are designed to have a specific chemical or biological activity, and this does not cease when they pass from the person, crop, animal or washing-machine which saw their primary use. Diluted, their effects may be much slighter, but the potential remains. An important property of such materials in the environment is their degradability, or its reciprocal, their persistence. This determines how long they retain their potency for side-effects. Some pesticides like DDT and other organochlorine compounds (such as dieldrin, aldrin or benzene hexachloride) owe their value in insect control to their great toxicity to insects, but lesser toxicity to other forms of life, in some cases coupled with their ability to remain in the environment in an active form for long periods. Substantial insect control can thus be achieved with a single application – an asset when seeking to control malaria in tropical swamps that have to be sprayed, at considerable cost, from the air. But many of the persistent pesticides share with metals the property of bioaccumulation, so that they can pass from low levels in the medium to high levels in organisms that stand on pathways and food chains. Some detergents are likewise persistent or 'hard' and cause rivers to foam in an unsightly fashion although they are relatively harmless to aquatic life. Other pesticides, detergents and drugs are 'soft' or biodegradable, or just unstable chemically so that they only persist for a short period. More than 90% of the anionic detergents which are the commonest in use today are biodegradable. Many new pesticides, like the organophosphorus group, break down readily in water (although not necessarily completely to simple inorganic residues). So do the 'nerve gases' whose dumping in the sea caused such a furore in 1971. Even if a substance is biodegradable, this does not mean that it poses no problems; the products of breakdown and the chemical transformations in the medium also need consideration, and this matter is discussed further in Chapter 3, which deals with pathways.

Oil is another troublesome pollutant of rivers, and the mounting problem it poses inland in many countries has been partly obscured by the emphasis on its role as a marine pollutant in recent years. Inevitably, it is at sea, the greatest sink for the world's persistent wastes, that most waterborne pollutants accumulate (FAO, 1971). There is continual deposition in the sediments of the ocean floor, and in this way considerable quantities of heavy metals and other substances are removed each year from the biogeochemical cycles. But persistent pesticides, polychlorinated biphenyls (a related persistent chlorinated organic substance also liable to bioaccumulation), and mercury, cadmium and lead salts are all present in the tissues of certain marine organisms, in some places at levels which make consumption by humans unwise. Oil has caused concern because of the disamenity of its presence on beaches, because it has killed many seabirds (Chapter 5) and because the methods used to disperse it have not themselves been harmless.

The Group of Experts on Scientific Aspects of Marine Pollution (GESAMP) has drawn up a 'black-list' (Category I) and a 'grey-list' (Category II) of pollutants considered most troublesome at sea. Table 7 is a summary, taken from the third report of the Group, setting out the

Table 7. *Major classes of marine pollutant and their effects*

Type of pollution	Harm to living resources	Hazards to human health	Hindrance to maritime activities	Reduction of amenities
Domestic sewage including food-processing wastes	+ +	+ +	(+)	+ +
Pesticides				
Organochlorine compounds	+ +	(+)	−	−
Organophosphorus compounds	+	+	−	−
Carbamate compounds	+	(+)	−	−
Herbicides	+	(+)	−	−
Mercurial compounds	+ +	+ +	−	−
Miscellaneous metal-based compounds	+	+	−	−
PCBs	+	(+)	−	−
Inorganic wastes				
Acids and alkalis	(+)	−	+	−
Nutrients and ammonia	(+)	(+)	−	(+)
Cyanide	(+)	(+)	−	−
Sulphite	+	−	−	(+)
Titanium dioxide wastes	(+)	−	−	(+)
Mercury	+ +	+ +	−	−
Lead	+	+	−	−
Copper	+	?	−	−
Zinc	+	−	−	−
Chromium	+	?	−	−
Cadmium	+	?	−	−
Arsenic	+	?	−	−
Radioactive materials	−	+	−	−
Oil and oil dispersants	+	?	+	+ +
Petrochemicals and organic chemicals				
Aromatic solvents	+	?	−	(+)
Aliphatic solvents	+	?	−	(+)
Phenols	+	+	−	(+)
Plastic intermediates and by-products	+	?	−	−
Plastics	(+)	−	+	+
Amines	+	?	−	−
Polycyclic aromatics	+	+	−	−
Organic wastes including pulp and paper wastes	+ +	?	(+)	+
Military wastes	+	?	+	?
Heat	+	−	−	−
Detergents	+	−	−	(+)
Solid objects	+	−	+	+ +
Dredging spoil and inert wastes	+	−	+	+

From GESAMP (1971).
Key to symbols: + +, important; +, significant; (+), slight; ?, uncertain; −, negligible.

main kinds of marine pollutant and the hazards they pose, while Table 8 analyses their sources. The full list of Category I materials contains 167 substances while Category II, judged less hazardous, has 229 entries. GESAMP's summary list is the basis of the lists of substances whose dumping at sea is prohibited under the Oslo and London Conventions and whose discharge from the land via rivers, pipelines and coastal outfalls is regulated under the Paris Convention (Chapter 9).

2.5.3 *Pollutants of food*

Fungi and bacteria often colonise grain, rice, potatoes and other crops stored as food for man and animals. Some of them have been known for a long time to make the food unfit to eat – generally because unpleasant symptoms have followed when uncautious or desperate people have consumed materials infected in this way. Ergot-infested grain, for example, has been known for centuries to cause abortion, gangrene, convulsions and death. In the 1940s widespread fatal haemorrhages occurred in the Soviet Union after people ate bread made from grain stored through the winter under the snow, and there infected with fungi. More recently, various outbreaks leading to the death of livestock have been traced to fungal toxins in their food (Austwick, 1975). It is now clear that mycotoxins are capable not only of causing acute illness and death, but chronic damage because they are carcinogenic and can also be teratogenic (i.e. produce embryonic abnormalities). There is evidence that they may be among the most serious, if not the most serious, pollutants affecting man and his livestock in developing countries, where the more familiar and more widely publicised pollutants of the general environment are of little significance.

Other substances can also contaminate food accidentally or as a result of deliberate human action, and examples are given in Section 3.2.4. Metals, nitrates, oxalates, nitrosamines, various organic acids, sorbic acid and sulphur dioxide are among the substances involved.

2.5.4 *Pollutant substances: consensus of views*

Despite the diversity of approaches and classifications, a comparison of evaluations undertaken by many different expert groups suggests that there is broad agreement as to the substances that present greatest hazard in the environment. Table 9 draws on lists prepared by the United Nations Environment Programme, the Economic Commission for Europe, the World Health Organisation, the Organisation for Economic Co-operation and Development and many national groups and presents this agreement in the form of a 'league table'. The table will undoubtedly change as

Table 8. *Principal sources of marine pollution*

Type of pollutant	Manufacture and use of industrial products – disposal via direct outfalls and rivers	Domestic wastes. Disposal via direct outfalls and rivers	Agriculture, forestry, public health via run-off from land	Deliberate dumping from ships	Operational discharge from ships in course of duties	Accidental release from ships and submarine pipelines	Exploitation of sea bed mineral resources	Military activities	Transfer from the atmosphere
Domestic sewage including food-processing wastes	+	++	-	+	(+)	-	-	-	-
Pesticides									
Organochlorine compounds	+	+	++	(+)	-	o	-	?	++
Organophosphorus compounds	+	(+)	+	-	-	o	-	?	+
Carbamate compounds	+	-	(+)	-	-	o	-	-	-
Herbicides	+	(+)	+	-	-	o	-	+	+
Mercurial compounds	+	-	++	-	-	o	-	?	?
Miscellaneous metal-containing compounds	+	(+)	(+)	-	-	o	-	-	?
PCBs	++	(+)	-	(+)	-	-	?	-	+
Inorganic wastes									
Acids and Alkalis	+	-	-	+	-	+	-	-	-
Sulphite	+	-	-	+	-	-	-	-	(+)
Titanium dioxide wastes	o	-	-	o	-	-	-	-	-
Mercury	++	+	-	+	-	o	-	?	++

Chromium	—	?	—	o	—	?	—	—	++
Cadmium	—	?	—	o	—	—	—	—	++
Arsenic	?	?	—	o	—	+	(+)	—	+
Radioactive materials	*⁄o	(+)	—	o	—	(+)	—	—	++
Oil and oil dispersants	—	+	+	+	+	+	—	(+)	++
Petrochemicals and organic chemicals									
Aromatic solvents	?	?	—	(+)	—	(+)	—	—	++
Aliphatic solvents	?	?	—	(+)	—	(+)	—	—	+
Plastic intermediates and by-products	?	—	—	(+)	—	+	—	—	++
Phenols	—	(+)	—	o	—	+	(+)	(+)	++
Amines	—	—	—	o	—	(+)	—	—	+
Polycyclic aromatics	—	—	?	o	?	+	+	++	++
Organic wastes including pulp and paper wastes	—	?	—	—	—	+	—	—	++
Military wastes	—	—	?	?	?	?	—	—	?
Heat	—	—	—	—	—	—	—	?	++
Detergents	—	—	—	—	—	—	(+)	++	+
Solid objects	—	+	(+)	(+)	++	++	—	+	+
Dredging spoil and inert wastes	—	—	++	—	—	+	—	—	+

From GESAMP (1971).
Key to Symbols: ++, important; +, significant; (+), slight; ?, uncertain; −, negligible; o, potentially harmful; *, dependent on extent of weapons testing.

Table 9. *'League table' of substances generally agreed to pose a hazard in the environment*

Division one

Air pollution

Sulphur compounds, especially sulphur oxides and aerosol sulphuric acid
Suspended particulates (the size is critical) (includes smoke and metal particles)
Asbestos fibres (and possibly other fibres of critical size)
Carbon dioxide (as a potential modifier of climate)
Chlorofluorocarbons[a]
Nitrogen oxides (as potential modifiers of stratospheric ozone and participants in reactions leading to photochemical smog)
Hydrocarbons (involved in photochemical smog reactions)
Oxidants (ozone and peroxy-acetyl nitrate or PAN)

Land

Radionuclides
Metals, especially cadmium, mercury and lead (according to formulation)
Persistent organochlorine pesticide residues

Fresh-water and marine pollution

Metals, especially cadmium, mercury and lead (according to chemical formulation)
Phosphorus (as an agent in creating excessive plant growth)
Nitrate and nitrite (both as agents in promoting excessive plant growth and directly as a problem in drinking water)
Radionuclides
Polychlorinated biphenyls (PCBs)
Oils and hydrocarbons (various forms)
Persistent organochlorine pesticide residues

Pollution in the domestic and industrial environment

Vinyl chloride monomer
Mycotoxins
Asbestos
Microbial contaminants[b]
Nitrosamines
Carcinogenic, mutagenic and teratogenic substances generally[c]

Division two[d]

Carbon monoxide[e]	Carbamates	Selenium
Fluorides	Organophosphorus pesticides	Titanium
Odours	Organosilicon compounds	Vanadium
Arsenic	Organostannic compounds	Nickel
Cyanide	Ammonia	Beryllium
Solids suspended in water	Carbon disulphide	Chromium
Surfactants (detergents)	Antimony	

[a] Chlorofluorocarbons (including those sold under the trade name 'freons') have attracted attention since many of the lists used in preparing this table were compiled, but would today probably feature in most entries (see DOE, 1976a).

substances are promoted and relegated in the light of new knowledge, improved control, or diminished use. New entrants whose possible hazard has been overlooked may be added. But it is safe to say that any major emission of 'division 1' substances is likely to cause real concern nationally and internationally today. It is also clear that 'division 1' is chemically heterogeneous, partly because it groups some materials of low toxicity that are being released to the environment on a very large scale alongside others of high toxicity but very much lower rate of discharge. Some chemical groupings do, however, emerge as warranting especial care: among them are organometal and organohalogen compounds.

2.6 *Properties of pollutant substances*

The basic rationale behind all these lists is straightforward. Whatever the classification used to sort out the mass of potential pollutants, those most likely to be troublesome have certain well-marked features. When assessing the problems a substance may pose it is, in particular, desirable to evaluate its:

(*a*) short- and long-term toxicity (which may also demand assessment of its biochemical reactivity, and the precise way in which it operates);

(*b*) persistence (and the ways and rates of its breakdown by both physical and biological means);

(*c*) dispersion properties;

(*d*) chemical reactions and breakdown or interaction products (which in turn may need screening for the other properties in the list);

(*e*) tendency to bioaccumulation;

(*f*) ease of control.

---

[b] Microbial contamination is not included in all lists because most authors treat this kind of contamination as different from chemical pollution. But bacterial contamination of food, stored products and potable waters remains a major environmental problem in many areas.

[c] Carcinogens etc. are very variable in nature, potency and levels in the environment and in practice some will fall in this division and some in division two. A very large number of substances can be carcinogenic if presented experimentally in very large doses: some will be in neither list because they are not encountered in significant quantities in the real world.

[d] Division two could be a very long and diverse list: only substances in common to several expert evaluations are included here.

[e] Carbon monoxide is placed in division one by many authors, but in practice it poses a health hazard largely in enclosed spaces and to people with respiratory disorders: environmental contamination rarely raises its levels in blood to values approaching those which habitual smokers derive from their tobacco.

Some of these factors overlap, and they are by no means independent or comparable variables.

Persistent substances of high toxicity which are difficult to control by technology have the capacity to create serious problems. Materials that disperse readily and are of low persistence are generally less troublesome because they will rapidly disappear from a particular place, even if highly toxic (this is why properly used organophosphorus pesticides are not a major pollution hazard). Materials that disperse readily yet accumulate within the tissues of living creatures and can be passed from prey to predator may pose a hazard even in conditions of good dilution. If synthetic substances with unknown long-term effects prove to be liable to bioaccumulation, there may be good grounds for caution.

Toxicity needs to be assessed in a way that will relate dose both to the targets affected (and their value to man and role in ecosystems) and to the circumstances of exposure. Toxicity (like dose) is related to concentration and to the time of exposure: a high concentration may kill in a short period, a lower concentration be lethal after prolonged exposure, while a lower level still may affect behaviour or susceptibility to environmental stress only cumulatively over the years (chronic exposure). The effects of some substances may only become manifest a considerable time after exposure: this is true of certain carcinogens and mutagens (substances producing cancer and genetic changes). Criteria (a quantitative statement of such relationships) are consequently difficult to determine.

Experimental assessments of acute (i.e. short-term) toxicity are relatively straightforward, but require controlled conditions in which a known concentration of pollutant is sustained for a predetermined period during which all other significant environmental factors like temperature are kept as constant as possible. The target organisms need to be chosen carefully. Certain species of known sensitivity that are also easy to keep in the laboratory are commonly used (e.g. trout and prawns for fresh-water and marine pollutants and alfalfa for air pollutants potentially toxic to plants) (Saunders, 1976). So far as possible the plants and animals are chosen to be genetically uniform, at the same stage of the life cycle, and having experienced the same conditions (including freedom from pollutants) over the period prior to testing.

Acute toxicity tests are often used to determine LC_{50} (or LD_{50}) values: these represent the minimum concentration (or dose) of the pollutant lethal to 50% of the test organisms in the period of the experiment (usually 24, 36, 48 or 72 hours, but sometimes up to 3 weeks). Other variants similarly determine estimated doses (ED) for a given damage rate short of death, or threshold limits (TL) defined as the combination of concentration and duration of exposure required to cause the first perceptible damage, for example by air pollutants to plant leaves.

Such tests are satisfactory starting points, but are not wholly reliable as indicators of the likely effects of exposure in the variable conditions of the real world, with many interacting factors and organisms at all stages of their life cycles. If the tests are to be used as an indication of tolerable concentrations of a substance in the environment, moreover, the selection of the test organisms is critical. Extrapolation from rats to man or shrimps to estuarine ecosystems is a process demanding care. Acute toxicity tests, finally, are unlikely to prove satisfactory as a means for evaluating the effects of chronic exposures. More sophisticated studies are needed to evaluate subtle biochemical and genetic effects and determine the likelihood of carcinogenesis, teratogenesis and mutagenesis, especially after periods of delay between exposure and reaponse.

There are two kinds of problem in interpreting such tests. One is statistical. In man we may be concerned with effects that only manifest themselves 20–50 years after initial exposure to a chemical, or may produce effects on children some years after the exposure of the parents. But a test that needs 50 years (and may then require repeating for verification) is of little use: we cannot wait that long. Some means of compressing the time scale is essential. Using a shorter-lived animal is one way, if we assume that in such species the latent period (gap between exposure and response) is also shortened. Increasing the dose is another way, if it is assumed that the proportion of a population affected can be related to the dose and that for any individual the effects are 'all or none': either cancer, mutation or embryonic abnormalities are induced or there is no effect. If this is so, we should be able to derive a model from exposures over a range of high doses which can be used to calculate the likely situation at lower doses.

The second problem is biological. Man, rats, mice, shrimps and bacteria differ in their responses to many substances. There is a recent trend to use tissues or micro-organisms in culture to test chemicals, especially for mutagenesis (one of the best-known and widest-used of these methods is the Ames Test). These methods are valuable because so many more generations can be examined in a short period than is possible even in an r-strategist mammal like a mouse, and the use of such cultures is also ethically preferable to killing large numbers of more advanced forms of life. But bacterial genes are unlikely to respond to many chemicals in the same way as human genes. Bacterial tests are hence useful chiefly as a first screen, indicating materials that need more thorough evaluation. Human tissues – or perhaps tissues from other long-lived K-strategists – in culture may be more useful in future (Peto, 1978). There may also be a use for tests on whole organs, maintained *in vitro* in the laboratory. All these tests need to be looked on as complementing epidemiological studies in the field and whole-animal laboratory experiments that will remain the most effective way of exploring detailed mechanisms of cause and effect for the forseeable future.

Persistence also need to be studied realistically, under conditions bearing some relationship to the natural world. For this reason, controlled laboratory tests need to be extended by ecological measurements. Most pesticides are now tested for their persistence in the soil and in plants grown on experimentally treated trial plots, followed over several years. Miniature environments in the laboratory also provide a useful scaled-down test-bed. For example, tanks have been set up with a bed of sloping sand on which cabbages growing at one end provide caterpillars with leaves treated with a pesticide: when the caterpillars die they fall into, or are transferred to, a pool at the other end of the tank where their remains are consumed by mosquito larvae that are in turn eaten by fish. The passage of a pesticide or its residues and breakdown products can thus be followed through a chain of organisms and the effects observed. Rather comparable continuous-flow systems, in which water of known composition and bearing known loads of pollutant is passed through tanks containing fresh-water plants or animals, and the uptake, tissue distribution, elimination and effects of the pollutant are studied over long periods are also likely to prove increasingly useful in evaluating the persistence and behaviour of substances in the environment, although these laboratory studies, like toxicity tests, need to be accompanied by field investigations.

Dispersion properties clearly depend on whether the pollutant is a gas, liquid or solid, discharged to air, water or soil. Diffusion coefficients, solubility in water or oil, and sedimentation rates or settlement rates from air and water currents of varying speed will all need to be considered. Here, however, the physical laws governing performance are well known, and realistic models of pollutant behaviour can be constructed. This is fortunate, for the actual behaviour of a substance in the environment clearly determines how far its potential toxicity can be realised. For example, it has been shown that DDT could inhibit photosynthesis in the planktonic plants floating in the surface layers of the sea and should this happen there would obviously be a major threat both to oceanic ecosystems and to global oxygen and carbon cycles. But it has also been shown that the solubility of DDT and its derivative DDE (the form normally present in the environment) in water is so low that the concentrations needed for such an effect are most unlikely to be attained.

Reactions and breakdown products must also be considered. For if a substance turns into something else in the environment, with different toxicity and dispersion, and different tendencies to accumulate in organisms, we are living in a fool's paradise if we look only at the properties of the original material. Moreover, we know that these transformations can be numerous and important. Figure 7 illustrates, in simple form, just a few that can occur in the atmosphere in photochemical smog formation. Inorganic salts of mercury can be transformed by bacteria in lake or estuarine sediments into the far more troublesome methyl

NO_2 Nitrogen dioxide	+	Light	$\longrightarrow$	NO Nitric oxide	+	O Atomic oxygen
O	+	O_2 Molecular oxygen	$\longrightarrow$	O_3 Ozone		
O_3	+	NO	$\longrightarrow$	NO_2	+	O_2
O	+	Hc Hydrocarbon	$\longrightarrow$	HcO· Radical		
HcO··	+	O_2	$\longrightarrow$	HcO_3· Radical		
HcO_3·	+	Hc	$\longrightarrow$	Aldehydes, ketones etc.		
HcO_3·	+	NO	$\longrightarrow$	HcO_2· Radical	+	NO_2
HcO_3·	+	O_2	$\longrightarrow$	O_3	+	HcO_2·
HcO_x· Radical	+	NO_2	$\longrightarrow$	Peroxyacyl nitrates		

Figure 7. Simplified reaction scheme for photochemical smog. This scheme is illustrative, not definitive; research is still in progress on the detailed chemistry of the smog-forming process. From American Chemical Society (1969).

mercury, which can then be accumulated in organisms. DDT breaks down to DDE, with subtly different environmental properties. Sulphur dioxide undergoes a variety of transformations which affect its impact on targets.

Tendency to bioaccumulation is often related to persistence, but not in a precise manner. It is probably a general rule that substances that are labile in the environment do not accumulate in living tissues. Those that do fall into two main groups. These are first, fat-soluble materials, especially halogenated organic molecules like the organochlorine pesticides and the series of polychlorinated biphenyls, which tend to be deposited in animal body fat reserves and mobilised back into the blood when the animal is starved, and second, metallic molecules present in the environment as particles or in organic combinations. Chemical and physical properties may provide reasonable grounds for suspecting that a substance will accumulate in tissues: tests can be done either by feeding to birds and mammals and sampling tissue or using filter-feeding invertebrates like mussels, known to concentrate metals and other substances present in great dilution in their environment. This property is the basis of a 'mussel watch' programme to monitor levels of bioaccumulated persistent pollutants around the world, recently supported by the United States National Committee for Problems of the Environment (Goldberg, 1975).

Scale of manufacture and ease of control must constantly be borne in mind if assessments are to be realistic. There is little point in worrying about the acute effects of most materials that are made in minute amounts (although if they are used as food additives or pharmaceuticals, or may be present as trace contaminants in water their role as potential carcinogens, teratogens or mutagens may need to be considered).

Likewise, substances used under constant surveillance, under conditions which prevent environmental contamination, may need attention in the factory but not more widely. The effectiveness of the containment does, however, need monitoring. For example, polychlorinated biphenyls caused widespread concern in the 1960s and early 1970s, when their effects on marine ecosystems, and especially sea birds became apparent (Chapter 5). As a result, in most countries they are now manufactured and used under conditions that should make environmental contamination impossible. Assuming that this policy is effective, their levels in the environment and seriousness as pollutants can be expected to decline, but monitoring suggests that up to 1976 there was no significant reduction in the residues contained in wildlife. Clearly, under such circumstances continued study is needed, and this should explore the possibility, proposed by Maugh (1973) that PCBs may be produced naturally through the oxidation of organochlorine pesticides over the oceans, using sunlight as the energy source: if this is happening, the anomaly may be explained and a new balance between policies for the control of PCBs and organochlorines may need to be struck.

The problem of accident will remain with us. However good our routines for containing a hazardous material in the factory, or for developing safe codes of practice for its use may be, we must also assess the likelihood and consequences of something going wrong. In Iraq in 1972, the distribution of seed corn treated with mercurial fungicide led to a disaster when famine-stricken people ate it (Bakir *et al.*, 1973). In 1973 a simple human error led to some 2000 lb of polybrominated biphenyls sold as a fire retardant being mixed into cattle food and sold widely to farms in Michigan, leading to massive losses of livestock, considerable human symptoms especially among farmers, and detectable levels in the serum and tissues of some nine million people (Selikoff, 1978). Complete proof against error and foolishness is impossible (as the well-known statement of 'Murphy's Law' concerning the ingenuity of fools stresses) but it is at least reasonable to assess what would happen if a substance 'got out' and what the cost of containing it with various degrees of certainty would be, hence relating vigilance to hazard.

2.7 Prediction of pollution hazard

2.7.1 Pointers to a problem

It is apparent that, in general, we can expect a substance to be a serious pollutant if, under actual circumstances of production and use it:

(*a*) has significant biological effects at low concentrations;
(*b*) diffuses readily in air, is soluble in water, or has a particular tendency

to accumulate in living tissues (such materials are commonly fat-soluble);

(c) is persistent;

(d) has breakdown products or combination products whose toxicity, persistence and capacity to reach or accumulate in targets equals or exceeds that of the original material;

(e) affects a wide range of organisms, and especially those important directly to man or central to the stability of global ecological systems;

(f) is produced on a large scale, in the course of operations which in all other respects confer major benefits on society (so that a large cost due to pollution may be acceptable in cost-benefit equations).

2.7.2 Screening

What emerges from this survey is that an almost infinite number of substances have some properties which could make them pollutants. Almost any substance can be toxic and ecologically disruptive at a sufficient concentration. Testing needs to assess the width of the margin between the actual circumstances of use and exposure and the dose at which unacceptable effects would begin.

The properties to screen for depend largely on circumstances of proposed use. If a substance is intended as a food, food additive or drug, its effects on man or livestock must be of prime concern, and tests will be needed on a range of mammal tissues over a long period of time (Lloyd and Drake, 1975; Spicer, 1975). Cosmetics obviously need scrutiny for their rate of absorption through the skin and effects on its cells. If a new pesticide is proposed, its toxicity to the organisms it is intended to control, its hazard to the human user (or bystander) and its side-effects on non-target organisms (whether crops, livestock, beneficial birds and insects or ecological systems generally) need evaluation. If the substance is a detergent, its breakdown in the environment and its effects on organisms in sewage works, rivers and seas must receive attention. Industrial chemicals need screening both to ensure the safety of those working where they are made or used and the safety of people, livestock, crops and wildlife and the protection of structures in areas where they may be released. Common sense should allow diagnosis fairly readily. But a codified procedure will help in:

(a) ensuring that all such materials are considered;

(b) determining that the right questions are asked;

(c) relating controls to the results.

One of the most recent national attempts to impose such a screening system is the United States Toxic Substances Control Act of 1976 (TSCA). This imposes a responsibility on manufacturers or producers of chemicals to provide the Environment Protection Agency (EPA) with adequate information about their behaviour and effects. The EPA must be notified and supplied with this information 90 days before a new chemical is manufactured commercially, and can limit manufacture, processing, use or disposal if it considers that a substance may pose an unacceptable risk.

Operation of such an act clearly demands very careful judgement. The number of substances potentially capable of creating risk is legion, and their molecular structure is a poor guide to hazard. The number of targets that could be affected is also immense. It is impossible to test everything for every potential effect on everything. Realistic simulation of plausible exposure levels will thus be vital. The United States already has legislation in the so-called Delany clause that prohibits the use in pharmaceuticals or food additives of any substance that has been shown to produce cancer in any mammal, and tests have amply demonstrated that if this is interpreted rigidly, irrespective of dose, many substances which display no known hazard under normal circumstances may have to be prohibited. This was the cause of recent bans on cyclamates and saccharin following development of cancers in some rodents fed massive doses. Care will obviously be needed to implement TSCA on a foundation of sound screening and risk assessment, and considerable problems are likely to arise in the process (EPA, 1977).

2.7.3 *National and international data banks*

Many actual and potential pollutants are released in a number of countries, and some are distributed widely in trade. It is clearly sensible on economic grounds as well as to avoid the creation of technical barriers to trade (Chapter 9) for testing to be harmonised and the resulting data made available internationally.

Various consolidated classifications are now being developed, as an essential framework for ordering the information in the referral systems, data banks and registers of toxic substances being set up by national Governmental bodies, regional organisations (including the European Communities) and international agencies under the co-ordination of the UN Environment Programme (Peachey, 1974). Details are given in Chapter 9.

The primary approach in most instances is by substance, with the chemical, physical and biological properties of each substance in the register entered in a fashion that permits retrieval of those that possess one or more specified attributes. Flexibility is clearly essential if the data bank is to be useful: it must have the capacity to answer many kinds of question. For example it may sometimes be useful to know which substances, manufactured in a particular quantity, exceed a specified acute toxicity to fish or rats: another enquiry, however, might relate to dispersal properties and demand knowledge of solubilities or volatility. One of the advantages of computer-held data banks is that multiple-attribute retrieval allows the information in the store to be re-grouped quickly according to a wide range of classifications and needs. Equally, such systems can be limited by time lags in the inclusion of information, and by constraints imposed by format and starting point. A vast amount of information is already available about materials already in use: is all this to be combed for inclusion in a data bank (in which case the effort and cost will be substantial) or is the bank to be restricted to new materials, in which case its use will be limited? These are the kinds of problem now being examined.

3. The significance of pathways

3.1 *Terminology: sources, pathways and sinks*

The concept of a pathway is fundamental in considering pollution. Pollutants enter the environment from somewhere – a cigarette end, a human bowel, a factory chimney, a tanker's bilge. This is the source. The scale of input clearly depends on the number and kind of sources in an area (this is one interaction with social factors). Within the environment, pollutants generally follow diverging pathways as they spread and are diluted. The rate of dispersion, and the distance a substance travels depends on its properties and those of the medium to which it is discharged. The main sectors of the environment – air, fresh water, soil and the sea – differ in their dispersion and dilution characteristics and this is one reason why it is sensible to consider them separately. Air mixes most rapidly and offers the prospect of very swift dilution and dispersion for gaseous emissions. Heavy particles or heavy gases, equally evidently, may settle out quite near to points of emission. When considering the dispersion of air pollutants therefore one is particularly concerned with the relative densities of the emitted materials and the characteristics of the receiving airflow.

Much the same applies to the sea or rivers. The sea has very great potential capacity to dilute and disperse wastes. What actually happens however depends on the solubility or miscibility with water and the diffusion characteristics of the substances discharged. Some materials quickly sediment out to the sea bed or river muds and may there be physically or chemically adsorbed and rendered inert. Living organisms may filter contaminants from air or water, and may then quickly break them down, or may alternatively accumulate them as long-lasting residues. Generalisation is impossible. What can be said, however, is that the concentration a pollutant attains at a point is the resultant of the quantity of input to the environment (from whatever pattern of sources, at whatever distance), the dispersion characteristics determined by the properties of the pollutant (density, solubility, diffusion coefficient) and those of the medium (current direction, rate of flow, rate of intermingling, adsorption properties) and the rate of removal from the environment at all points along the pathway, whether caused by physical or biological agencies.

44

It is immaterial how hazardous a substance may be in theory, if in the real world it is so rare or so safely contained that it never reaches a target. The whole thrust of control measures, when handling the most dangerous materials, should therefore be towards ensuring that targets are not exposed to them – not only under ideal operational conditions but under all reasonably probable circumstances, including accidents. Where emissions of a substance to the environment are tolerated, controls need to be adjusted so that targets are not unduly hazarded (just what constitutes undue hazard depends on the nature of the target and the value set upon it). It is impossible to establish meaningful controls if there is ignorance of how the pollutant can reach the target – that is, of the pathways it may follow.

Certain intrinsic properties that determine the likely effect of a pollutant were set out in Chapter 2. They fall into two groups. Some may be termed 'effect-generating properties' like toxicity or the capacity to react chemically with a surface and cause corrosion. Others are 'pathway-determining properties' like diffusion coefficients, solubilities or tendencies to bioaccumulation or biodegradation. These affect how far and how fast the substance spreads in the environment. Some, like solubility, may come into both categories. The scale and circumstances of generation of a substance are also relevant here, for it is the concentration of the material at the point of emission that determines the concentrations reached at varying distances away, through the operation of the normal laws of physics.

Figure 8 develops the pathway model in more detail. It is essentially linear: the amount reaching the target is a function of the amount emitted, diminished by dilution and removal and enhanced by biological or environmental accumulation factors. The distance and rate of movement along the pathway are critical, for these, combined with the rate of removal, and return to the medium, determine the degree to which material is attenuated by dilution and transformation. It is a simple transport system, in which except for anomalies such as bioaccumulation or accumulation in small and unusual regions of the environment, concentrations steadily diminish. Such anomalies of local accumulation can be looked on as closed loops from which material later re-enters the environment when an accumulator organism dies or sediment is re-mobilised. The feedback loops in the system arise when effects produced at the target modify transport or removal rates (as eutrophication affects the capacity of a lake ecosystem to oxidise further inputs of organic matter).

3.2 Pathways in the environment

The pathway from source to target has two main components: the portion (the greater part in terms of extent) lying in the air, water or soil and the

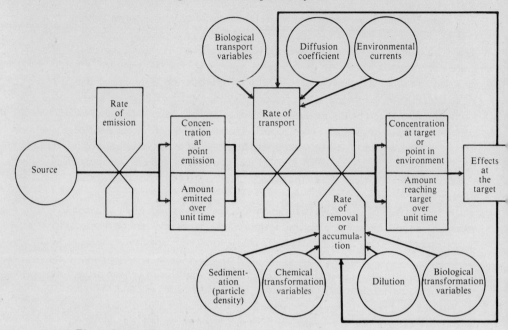

Figure 8. A pollutant pathway.

much shorter section, which none the less is of great importance, within the target, by which a pollutant is absorbed and transported to a point of impact – or excreted back to the environment. Within the environment, moreover, rates and mechanisms of transport in air, water and soil are very different. Commonly all are involved in the distribution of a substance (as Figure 9, a diagram of the pathways of sulphur, illustrates). But it is convenient to discuss the sectors of the environment separately here.

3.2.1 *Pathways in air*

The atmosphere has three main layers. Next to the surface of the earth is the *planetary boundary layer* – a thin layer of air, to which most pollution is discharged and which is greatly affected in its flow by the roughness of the land surface. This boundary layer merges upwards into and is really a part of, the lowermost main layer, the *troposphere*, in which the air gets cooler with increasing height, up to an altitude of around 10 km at the poles and 16 km at the equator. Above this level, the tropopause, there is a reversal in the temperature gradient and the *stratosphere* begins: it in turn ends with a second reversal of the temperature gradient at about 50 km. Above this lies the *mesosphere*: not a region of great concern in the consideration of pollution.

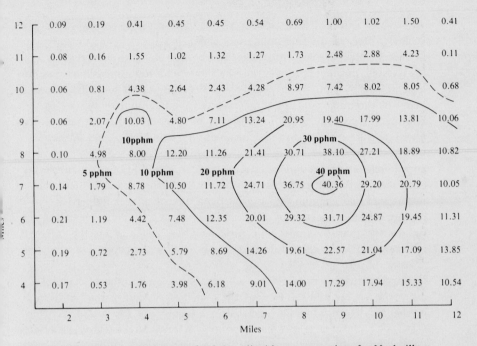

Figure 10. Computer calculations of sulphur dioxide concentrations for Nashville, Tennessee. Abscissa and ordinate are city dimensions in miles: isopleths ('contour lines') are concentration values in parts per 100 million (pphm). From *Man's impact on environment* by Detwyler. © McGraw-Hill Book Company. Used with permission of McGraw-Hill Book Company.

the ozone layer have attracted much attention in recent years (CIC, 1975; DOE 1976a; CISC, 1976). These materials are widely used in developed countries as refrigerants and aerosol propellants. They reach the stratosphere slowly, through the interchange between troposphere and stratosphere, because they appear hardly, if at all, broken down in the lower atmosphere. Their lifetimes in the air have been calculated at from 29 to 205 years, depending on the particular materials and on the assumptions made. In the stratosphere chlorofluorocarbons are destroyed by the more intense radiation, yielding chlorine and fluorine atoms and various free radicals and molecular fragments. The fluorine is expected to form hydrofluoric acid, which on downwards diffusion to the troposphere will be washed out in rain. The chlorine may react with ozone, reducing the amounts in the air, and also with methane, producing hydrochloric acid, some of which diffuses or is carried back into the troposphere and also washed out in rain, thereby providing at least a partial final sink for these materials. These degradation processes are so slow, however, and other minor sinks so small, that almost all the chlorofluorocarbons produced so far are probably still in the atmosphere, and observation confirms that this is likely. Because of this long time scale, and dependence on diffusion between stratosphere and troposphere for the ultimate removal of breakdown products, it will be seen that chlorofluorocarbons can be expected to be globally dispersed and exert any effect globally, whereas many of the impacts of the much shorter-lived sulphur oxides are in the boundary layer and local or regional in character. They thus clearly illustrate the contrasts that arise between short- and long-lived substances, with pathways involving different atmospheric layers, and confirm that even low levels of emission at stratospheric levels could be of much greater concern than the same emissions lower down. Rather similarly, radioactive inert gases like krypton-85, emitted from nuclear power industries, have no significant sinks except the atmosphere, and are only degraded by radioactive decay. Their concentration thus depends on the balance between emission, dispersion and this breakdown process (Dunster and Warner, 1970).

These pathways through the air determine how much of a pollutant arrives at a potential target. But the very last section of the pathway is crucial in determining the actual dose of pollutant the target receives. Gaseous pollutants reach plant, animal or structure surfaces at a rate depending on the resistance to diffusion within the boundary layer over the target surface, and this in turn depends on the rate and turbulence of the airflow (Ashenden and Mansfield, 1977). Many living surfaces are rough, with hairs or ridges, and these tend to slow the air and trap particles. Entry into plant or animal tissues commonly depends on solution in water at a surface from which the dissolved pollutant can move by diffusion in the liquid phase into the tissues, in cell fluids or blood (see Section 3.2.2). The last part of an atmospheric pathway is limited by the ease of flow of gases or suspended soluble particles through the respiratory passages of animals (dependent in vertebrates on the rate of lung ventilation and in insects on the ventilation of the internal breathing tubes and on whether the spiracles where these tubes reach the surface are open) or through the stomata that control entry to the spaces within the leaves of plants. Very little penetration of most pollutants, by contrast, is likely through the waterproof waxy surface layers of many plants and insects (Holdgate, 1979). In any event, the dynamics of this last part of the pathway are critical – and they are least well understood.

3.2.2 Pathways in water

Only a few gaseous pollutants (such as carbon dioxide or chlorofluoro-carbons) are accumulating in the air at the present time. Even for these the atmosphere is not a permanent repository: the carbon dioxide will eventually be taken up by biological cycles and by solution in the sea, finding an ultimate fate in marine sediments or new oil and coal, while the chlorofluorocarbons return to earth as inorganic chloride or fluoride. Gases with a more rapid removal, like sulphur dioxide, do not even accumulate in the air over a short time span: raised emissions are balanced by raised depositions, and it is the total amounts moving along the pathway that increase.

The pathways in fresh water are rather different, as Figure 11 shows. First, a lake, and even more so the sea, is a true 'sink' in which materials tend to accumulate. Inputs from natural erosion and man-made sources, unless taken up in biological cycles, have only one further sink open to them – sediments, returning materials to the lithosphere, the rocks of the earth, from which their constituents ultimately came.

Inputs to fresh water and sea come largely from the land, for even natural marine erosion is concentrated in the upper layers of water where the waves break. The material remains in the water column for a very variable period. At one extreme some is deposited almost at once as sediment on the ocean floor: at the other some dissolves and spreads by diffusion, and especially by currents. As in air, currents are much more important movers of material than physical diffusion processes. Particle size, density and solubility are key determinants of behaviour, governing the balance of sedimentation against transport.

Light, insoluble materials such as oils naturally float at the river, lake or sea surface and fat-soluble substances like DDT dissolve in the fatty or oily layers in such situations: heavy insoluble matter, equally naturally, sinks swiftly. Uptake by organisms depends on the form, and generally the solubility of the material. Some bacteria can, however, mobilise and make soluble inorganic salts otherwise incapable of entering the fresh-water system: this is why inorganic mercury, which presents relatively few problems in the environment, becomes a major concern when bacteria transform it to methyl mercury. Butler (in SCOPE, 1977) states that almost all the mercury present in an aquatic system is present in the sediments in inorganic form (the ratio of mercury in sediment to mercury in water is about 5000:1). The 'half-life' of this mercury in the sediments is from 1 to 5 years. Bacterial action converts this to the soluble methyl form at a rate of about 2 μg/(m^2 sediment day) (at 15 °C), and this dissolved mercury is accumulated by fish, which contain about 3000 times more methyl mercury than the water does.

As with gaseous pollutants, therefore, the circulatory properties of a water mass are critically important. There are lakes, river reaches, or estuaries which are virtually stagnant. Heavy pollution here is liable to have far more extreme effects than in well-mixed waters. Turbulent rivers, moreover, are naturally better oxygenated through the continual breaking

The significance of pathways

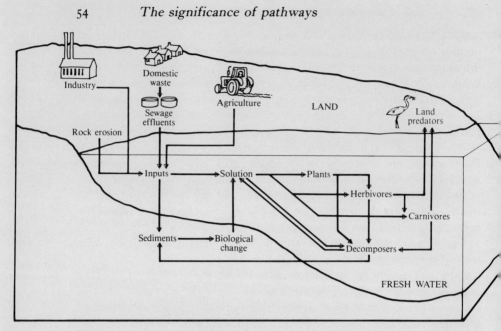

Figure 11. Pathways of pollution from land to fresh water, and within an aquatic system.

of the surface and 'bubbling' of air through water at little falls and rapids: even with a heavy input of inorganic matter they are less likely to become deoxygenated than a slow-flowing lowland stream. In estuaries, despite the apparent breadth and mixing with the ebb and flow of the tides, there is similarly a restriction on dilution, for the fresh water coming down rivers mixes only sparingly with the more saline waters to seaward, tending to form a layer floating upon the salty basal strata. Penned between banks, the water masses of an estuary surge up and down with the tide, and it may take many tidal cycles for a mass of incoming fresh water to pass out to sea. Estuaries, for these reasons, are relatively vulnerable to pollution (Porter, 1973; Royal Commission, 1972b).

Confined seas share this feature. Oslo Fiord and the Baltic Sea are well known to contain higher concentrations of many pollutants than the Irish or North Seas, which are in turn more contaminated than the open Atlantic. Inshore waters, likewise, are the places where bioaccumulation of pollutants is most likely, for many of the organisms involved are molluscs attached to the substratum in or just below the intertidal zone. Enhanced levels of cadmium have been demonstrated in shellfish in many estuaries contaminated by industrial discharges, with values of up to 60 ppm in extreme situations: commonly there are gradients from such high values in the immediate neighbourhood of emissions to figures below

10 ppm some kilometres away, and much less in unpolluted waters. Concentrations of mercury in fish from many inshore European and North American seas are in the range 0.5–1 ppm while in waters offshore values around 0.1–0.3 ppm are more normal and many oceanic species contain less than 0.05 ppm (MAFF, 1971; 1973).

3.2.3 *Pathways in soil*

Pollutants reach the soil in various ways: including deposition from air, either dry or in rain or sprays, deposition from flood or irrigation waters, deposition from living organisms, deposition from the sea in driven spray and through the dumping of solid wastes. Once on the soil surface, pollutants tend to be moved relatively slowly, in percolating drainage water or by organisms. The pathways tend therefore to be shorter, and contamination more local and patchy than in water or air.

Soils, like marine and fresh-water sediments, commonly retain pollutants because these are adsorbed onto mineral particles or otherwise trapped in relatively inert form. Only a small part of the lead or fluorine in soils, for example, appears to be available to plant roots. Normal soils have 2–200 ppm lead in them, but only 0.05–5.0 ppm is usually soluble (DOE, 1974b). Like mercury in lake silts, such metals may be released by biological action – or rendered inert by processes in the soil – but insufficient is known about these processes.

3.2.4 *Pathways modified by man*

The formation of the urban and industrial communities of the developed world has led to a substantial modification of many of these pathways. That involving the air has been least affected. None the less, some pollutants are potentially hazardous because they are presented to people in poorly ventilated air spaces including tunnels, underground car parks, and badly designed buildings. Carbon monoxide concentrations, for example, commonly reach 25–30 ppm in the restricted air circulation of town streets in rush hours, and higher levels are likely if there are traffic jams in tunnels: even in busy streets, however, the levels of carboxyhaemoglobin in the blood of people rarely exceed 3% of the saturation level, whereas smokers commonly exceed 5% and may exceed 15% (Lawther, 1975). The pathway of drinking water to man and farm livestock is substantially more modified, and may involve prolonged contact with lead pipes, in some circumstances leading to considerable increases in dissolved lead concentrations (see Section 4.5.2). Water treatment with chlorine, and sometimes ozone, in developed countries has virtually eliminated this pathway by which pathogenic bacteria formerly reached man, but the increasing re-use of water and the acceptance of rivers with a certain amount of pollution, for example with nitrate, metal salts and organic materials as sources of supply, has established routes by which small quantities of these materials can reach people and stock (Martin, 1975).

Food is an important pathway to man for many substances: it is, for example, the main route by which people in Britain take in lead. It must be stressed that many potentially hazardous substances are naturally present in food, and some are well known to cause occasional illness or even death. Rhubarb, for example, is rich in oxalates, cassava contains cyanide, and green potatoes form the alkaloid solanine: all can be dangerous. There is also a wide range of potential carcinogens in man's diet. When food becomes infested with certain moulds (as when it is stored badly) a range of toxins and carcinogens, including aflatoxins and nitrosamines, are often formed, and these are among the most widespread and hazardous of pollutants in some areas (Austwick, 1975). To these natural, or naturally produced, hazards (well reviewed by Crampton and Charlesworth, 1975), others have been added by man accidentally. Selikoff (1978) records the consequences of contamination of animal feed in Michigan with polybrominated biphenyls, which were subsequently passed on when people ate the meat and other produce. Yet other substances are added to food as a preservative, colouring or flavouring. Preservatives are of value because they have selective toxicity, controlling fungi or bacteria without (hopefully) harming man or other consumers of the foodstuff: because of this biological activity there is always the possibility of unwelcome side-effects and for this reason all such substances require very careful screening (Lloyd and Drake, 1975). Nitrite, for example, is a valuable additive to meats because it inhibits the organisms that form the poisons causing botulism: avoidance of this very real risk however involves a slight (and as yet unproven) one that it, or nitrosamines formed in the food or the gut of the person eating it, may produce cancer.

3.2.5 *Interlinkages between pathways*

The length of a pathway is generally related to the density and degree of confinement of the medium. Pollutants in air tend to have the longest pathways and to be best mixed and diluted. Pollution in the open sea is likewise mixed widely and diluted greatly. Land-locked seas, estuaries, rivers and lakes are increasingly prone to severe accumulation of pollutants. Soils are most likely of all to display high local concentrations. Conversely, the risk of spread from one site to another follows the opposite trend to that of concentration.

As Figure 9 indicated, pathways in air, water and soil interlink. Sulphur passes from air to drainage systems and the sea. Mercury from the land may pass to marine sediments, only to return to fresh waters in methyl form in the tissues of migratory fish. DDT and PCBs probably reach the sea through the air even though they have very low vapour pressures: it is this wide dispersion that accounts for their presence as measurable residues in the tissues of animals like Antarctic penguins, living far away from the regions where these materials are manufactured or used. Some of the oil in the sea also comes from the land, distributed as fine droplets in the air or borne down the rivers as an emulsion or surface contamination.

3.3 Pathways within targets

3.3.1 Pathways in ecosystems

Pathways in a medium are thus governed by the factors determining the transport, dilution and transformation of pollutants. They commonly overlie or lead to pathways within ecological systems, illustrated in Figure 12. Here the critical factors are those governing first, presentation to organisms; for example the levels in soil adjacent to roots, reaching foliage as gas, sediment or dissolved in rain, in the air or water in which the organism lives, or in food. The second group of critical factors are those governing uptake, which in turn depend on the chemical state of the material, including its solubility. The third group concerns the transmission of material from organism to organism in food chains, the fourth the inactivation of the material biologically (detoxification) and fifth, the recycling of the material via the decomposers or through other organisms capable of re-accumulation and re-conversion of the material to earlier states. The basic model is capable of much variation and some of the major variables are discussed in the following sections.

3.3.2 Pathways within organisms

One critical factor too commonly ignored is the fact that many pollutants are only partly absorbed into the body of an organism from food or air. For example, experiments in man show that only about 10% of the lead present in food or water is absorbed from the gut of adults into the tissues. The proportion may be as high as 50% in children, and naturally depends on the chemical form of the lead (DOE, 1974b). From the air, where the commonest form of lead contamination is as fine particles, alone or attached to larger aggregates of carbon, the uptake is more variable, but about 50% of the lead in the air inhaled is deposited in the lungs (the range is wide, from 10 to 90%). Half of this lead is now believed to pass into the blood, only a small proportion being brought up the respiratory passages by ciliary action and then swallowed. Only about half the lead

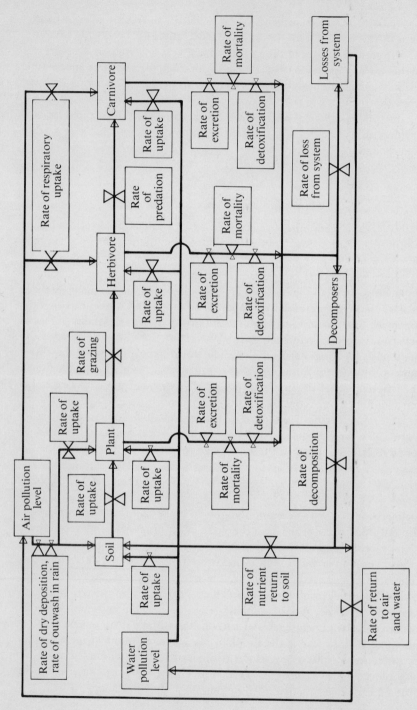

Figure 12. Pathways of pollutants and other substances within the ecosystem.

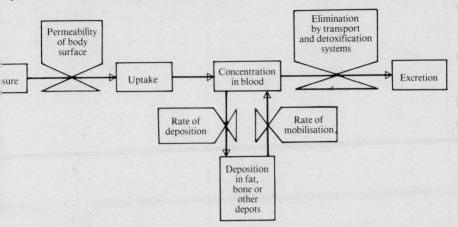

Figure 13. Factors controlling concentration of pollutant in an animal's blood.

in blood plasma becomes attached to red cells, the other half being removed to bone and other tissues. The basic model determining the level of a pollutant in the blood – the main factor in determining its biochemical effect – is shown in Figure 13. The first critical parameters are those determining its passage through the body wall, including gut wall or lung wall, and so entering the tissues. Some plants and animals are well protected against intrusion. Insects, for example, have tough outer layers or cuticles, waterproofed with wax that is shielded against mechanical disruption by a tough protein 'cement'. Their surfaces are not easily wetted by water: this is why a wetting agent or detergent is added to water-based insecticides, or such pesticides are oily, since oils wet most insect surfaces readily. Particles are generally 'safe' (although this is not true, for example, of asbestos and other materials in the mammalian lung) because plant and animal surfaces are not readily penetrated by them. The main sites of pollutant uptake are those specifically evolved for gaseous or chemical exchange: the stomata and leaf interstices of plants, the gills of fish, the lungs of mammals and birds, the spiracles and internal breathing tubes of insects and the guts of all animals. The skin of many mammals is also permeable to certain kinds of chemical solution, albeit far from freely.

These permeabilities influence the rate at which a substance enters an organism. Another factor is the concentration of the pollutant to which the organism is exposed. Chemical flow is along gradients: if there is a high concentration outside the organism, and a low one inside, there is likely to be more considerable intake than if the external concentration is very slight.

If uptake were simply a matter of chemical diffusion, the process would

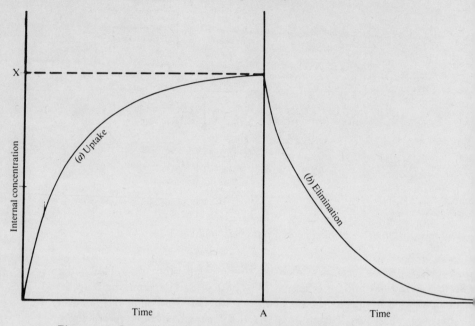

Time   A   Time

Figure 14. Generalised curves for uptake and elimination of a pollutant. X represents the level at which internal and external concentrations are equal: A a time at which external concentrations are reduced to zero.

in fact continue until external and internal concentrations were the same: the time required would depend on the permeability of the organism's surface, the initial relative concentrations and the temperature (since the transport process is temperature sensitive, going fastest in warm conditions). The system would thus follow a curve like Figure 14 and the plateau of internal concentration, at level X would be the same as the level in the medium.

In practice, almost all pollutants are also eliminated from living organisms. The processes vary. Some materials are excreted directly: most are first modified in some way to facilitate transport and outward passage through the body wall in lung, kidney or equivalent organ. Excretion is an active biochemical and physiological process and its rate, while tending to rise as the levels of a contaminant in the body rise, is less likely than absorption to be describable in simple diffusion terms. None the less, over time its effect is to reduce internal concentrations as in Figure 14b. If, in fact, the external concentration were reduced to zero at the time represented as A on the figure, the decline would follow a curve which is virtually the reverse of the uptake curve, though not necessarily on the same time scale. On an 'ideal' curve of this kind the amount in the body increases or declines by a constant factor over unit time – that is the

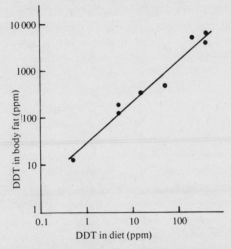

Figure 15. Concentrations of DDT in the fat of female rats fed DDT in their diets for up to 6 months. Data from W. J. Hayes: figure from Moriarty (1975b).

internal level is doubled, or halved, every so many days. In real life, however, both processes proceed simultaneously and the result is that the 'plateau' of concentration attained in the blood or tissues will be reduced below that otherwise inevitable: the plateau value depends on the balance between factors determining uptake and those determining excretion (Moriarty, 1975a, b).

Various experiments have confirmed the reality of this kind of pattern (e.g. Figure 15). However, conditions in nature tend to be more complicated. One variable is imposed by the fact that a pollutant may be removed from the bloodstream but deposited in bones, feathers and fat rather than excreted (feathers and hair do provide intermittent excretion when moulted). Bone lead can re-enter the blood (and chemicals can be provided to aid the process, as a medical technique designed to lower the body burden) but the rate is slow. Similarly, DDT, PCBs and other chlorinated organic compounds are deposited fairly slowly in the body fat of birds and mammals: from this depot they are more readily mobilised, especially when an animal uses these food reserves in times of scarcity. Starvation and stress may indeed flush them into the blood so rapidly that damage is caused to the nervous system or kidney, and death may result (Chapter 5). For obvious reasons great care is needed in clinical treatment designed to remove accumulated lead or other substances from bones or other depots in man. The levels of a substance in blood at any time thus reflect this interchange between depots and the bloodstream as well as the balance between intake and excretion.

The concentration of the *residue* of a pollutant in an organism varies greatly from tissue to tissue. The high levels in 'depot' tissues like body fat and bone have already been mentioned: much lower concentrations, for example of DDT and PCB, are normal in muscle or nerve tissue. If statements are to be made about the residue level of a substance in a target, it is therefore important either to give the result as whole-body burdens or to specify the tissues involved (see Table 16). Moreover, to interpret the biological significance of a residue level it is necessary to know the input and output rates involved: a high residue that is almost impossible to mobilise (like lead in bird feathers) may be of slight physiological significance whereas a much lower level in an active tissue may be a real indication of problems. The physiological condition of the individual is also important: a starved animal that is mobilising its body fat will have unusually high blood, liver and kidney concentrations of materials that are more normally concentrated in the fat reserves.

In some organisms the accumulation of metals, organochlorines and other substances to concentrations much (often a thousandfold) higher than in the adjacent medium may result from processes that have adaptive value. Sea squirts (*Ciona*), for example, use vanadium in their blood pigment just as man uses iron in haem, and it is consequently reasonable that they have the ability to concentrate this element from very dilute sea water. It is less easy to see any survival value in the accumulation of cadmium, lead and organochlorines by molluscs, cadmium by the marine insect *Halobates* or of mercury by some fish which appear to have natural levels as high as 1 ppm in muscles. These processes of accumulation reverse in some places and at some times, the general trend towards increasing dilution once a toxic material is released into the environment. Some of the accumulation, like that of organochlorines by marine plankton, appears to be a passive process: in other cases, as for those plankton's uptake of metals, it involves active metabolism. The rates of uptake of a substance by an organism naturally reflect both the physicochemical and biological mechanisms involved and the exposure conditions: a free-swimming organism like a fish may for this reason be less suitable as a sampling device, because there can be no certainty where it has been, than a sedentary mollusc sampling the water masses moving past it.

3.3.3 *Pathways in food chains*

Pathways between medium and organisms, and the uptake–output ratios which determine residue levels, naturally vary and account for different levels in different species and individuals. Sometimes residues also pass from organism to organism and from this has grown the well-known theory of pollutant concentration as it passes from prey to predator along a food chain. If a seed grain, an insect or an earthworm with a high content of a pollutant is eaten, the residues within the food may naturally affect the consumer, and if the diet of a species is high in a substance to which it is sensitive, there may be a significant, and even lethal effect. Seed corn dressed with dieldrin undoubtedly killed many pigeons in England a decade ago: the poisoned pigeons undoubtedly killed predators such as foxes (Mellanby, 1967). But the critical thing is not the total intake of toxic material from the food but the balance of intake and elimination, and if the predator can eliminate the toxic substance fast enough, and so keep its internal concentration low, it may suffer no ill effects. Jefferies and Davis (1968) have shown that in thrushes fed from 0.15 to 5.69 ppm dieldrin in earthworms for six weeks, only 0.8–4.4% of the material remained in the body at the end of the period and the concentration of dieldrin in the birds was *less* than that in the diet, suggesting effective elimination was sustained.

Moriarty (1975b) who has analysed this situation, suggests that the classic 'concentration along a food chain' story derived from Clear Lake in the United States is erroneous because the lower members of the food chain also had lower capacities to accumulate DDD, the insecticide applied to the lake to control mosquitoes. Later work demonstrated experimentally that a water plant (an alga), a crustacean (*Daphnia*) and a fish (the guppy, *Lebistes reticulatus*) differed in their uptake of dieldrin, *Daphnia* reaching a level about ten times and the fish forty times that of the alga, when all were placed in water of comparable concentration and allowed a comparable period to attain the plateau residue level. It follows that the fish, at the higher end of the food chain, contains more residue because it takes up more dieldrin, not because of concentration due to the passage of the pesticide along a food chain. To interpret the pattern of residues in the components of an ecosystem it is necessary to know the input:output ratios of the component members of the system, taking *all* forms of input into account, not only that from diet.

Butler (in SCOPE, 1977) has worked out such a situation for mercury in fresh water. His calculation suggests that both pathways – direct uptake from the water via the gills and intake from food organisms – are of about the same importance. The calculations also indicate that the proportion of mercury in the methyl form will increase with passage along a food chain. If the amount of mercury in a river sediment, the rate of conversion to the methyl form, and the flow rate of the river are known, Butler's equations permit calculation of the expected concentration in fish and other components of the system – and hence, since for man a most significant factor is the level in fish considered for human food – of the adequacy of control of mercury-bearing emissions. His example illustrates how important it is for controls to be designed with pathways in mind, so that they are directed realistically at the causes of genuine problems.

4. Changes in the environment

4.1 *The significance of environmental concentrations*

The preceding chapters have demonstrated the diversity of the substances that can act as pollutants, and the varied pathways they may follow. In evaluating how serious a particular pollutant may be, in addition to basic knowledge of its physical and chemical properties and of the routes by which it may reach a target, it is necessary to know the responses of targets to particular magnitudes of chemical or physical change in their environments.

For this reason, all studies of pollution are concerned to establish the concentration of potential pollutants in the media of air, water or soil (or the scale of perturbation of physical factors like temperature, sound or radiation). But, because of variation both in emissions and in factors like climate, that modify pathways, these exposures naturally vary from time to time. What matters to a target is the *dose* of a pollutant that it receives, and this is the product of exposure concentration and exposure time. If we can forecast how much pollution a target or receptor is likely to receive (a statement termed its *dose commitment* in radiation biology and sometimes more generally by scientists writing about pollution, for example Machta, 1976), and also know how the receptor reacts to the particular substance or factor, we should also be able to forecast the likely damage. *Harm commitment* is a term sometimes used to define the damage a target or target population is likely to sustain from a particular dose commitment, and the relationships between dose and harm, or exposure and effect, are sometimes termed *criteria* (see Chapter 5). Clearly, the ability to determine how far exposures, doses and damage are changing, and are likely to change, as a result of human actions is crucial to our pollution control policy.

Consequently, measurement of the magnitude of chemical or physical change in the environment is a major preoccupation of agencies responsible for pollution control, and Chapter 7 describes the many kinds of monitoring scheme. In the present chapter the nature of the various physical and chemical changes so far detected in the different sectors of the environment will be described, together with their trends. There is a vast literature about levels and trends of pollution in various media,

64

locally, nationally and globally, and total coverage is impossible: the references in these pages are illustrative rather than comprehensive. For good general accounts of the British situation the best starting points are two reports of the Royal Commission on Environmental Pollution (1971, 1974). The United States position is reviewed in the annual reports of the Council on Environmental Quality (CEQ, 1970–6), while global issues have been discussed by SCEP (1970), Institute of Ecology (1971) and SCOPE (1977).

Changes in physical or chemical environmental factors alter the quality of the environment as a habitat and in a sense imply that the media of air, water or soil act as 'targets'. The analogy is close where the medium is substantially altered – for example when temperatures are raised, currents disturbed or biogeochemical cycles modified. But the parallel is less complete when the change simply involves alterations in the concentration of component or dissolved substances without evident change other than in the organisms or structures affected by that substance. In this chapter the media are not treated as true targets but as components of the habitat whose features are modified to the detriment (or benefit) of targets within them.

Generally speaking, knowledge of the mechanisms by which various pollutants can alter physical factors or accumulate in parts of the biosphere is more complete than knowledge of the magnitude of the changes going on in various places. There is considerable contradiction between different books and reports. Much of the popular literature about pollution implies that it is getting worse (see, for example, Taylor, 1970; Ehrlich and Ehrlich, 1970; Meadows *et al.*, 1972; Commoner, 1972 or Tucker, 1972). Other books (e.g. Maddox, 1972) and reports, especially from Governments and scientific agencies, take a more optimistic line (e.g. Tukey, 1965; SCEP, 1970; Royal Commission, 1971). Are we 'winning' or 'losing'? Such a question can only be answered definitively if the present patterns of pollutant emission, pollutant dispersal, pollutant removal from the biosphere and concentrations over the world as a whole are known, and there is information about how these patterns are changing. Uncertainty arises because monitoring networks are localised and imperfectly designed (Chapter 7). Few surveys have defined the variations in conditions over a single country or region, let alone the whole planet. The general picture is compounded of guesswork and extrapolation, and this is largely why there is room for doubt about the justification for gloom or optimism. For if the known worst cases – the loss in oxygen levels in the lower Great Lakes, Lake Maclaren and the Baltic, the industrial pollution in the Rhine, the mercury levels in Minamata Bay, the soil contamination around Seveso, the PCB and DDE levels in Swedish, British and American coastal birds, the sulphur oxide and smoke levels in the London of the 1950s, the oxidant smog of Los Angeles – are generalised and it is assumed that urban and industrial growth must mean the coalescence of such 'hot spots' into a torrid globe, then indeed the prospect is very bad indeed. But extrapolation is a dangerous thing: clearly it is not intellectually honest to choose such worst cases, applying to a fraction of a per cent of the earth's surface and treat them as universal (Ashby, 1973a, b, 1975). Nor is it valid to select examples solely from the other extreme in the pattern of variation and claim that the world is perfect. In the following pages the fragmentary evidence that we had for current level and trends, globally and in Britain in 1976 are examined and their possible meaning evaluated. It is more than likely that within a few years new figures will have altered part of the picture.

4.2 Changes in the media

The principal effects of pollution on the media are:

(*a*) Physical effects on the air:
  (i) temperature changes;
  (ii) effects of noise and vibration;
  (iii) effects of gases and particulates on thermal transmission;
  (iv) effects on ozone layers and hence on ultraviolet radiation penetration;
(*b*) Physical effects on water:
  (i) temperature changes (some leading to changes in currents);
(*c*) Chemical effects on the air:
  (i) changes in the composition of the atmosphere;
(*d*) Chemical effects on water:
  (i) effects of oil at the air–water interface;
  (ii) changes in the concentration of dissolved and suspended substances.

These vary in scale and certainty. The most publicised global effects are undoubtedly those of gases and particulates on the radiation transmission and heat balance of the atmosphere, while at local level changes in temperature and chemical composition of air and water are widely reported.

4.3 Physical effects

4.3.1 Heating the air

It has been known for many decades that the air over cities is warmer than in the country around, due to waste heat emissions, mainly from fuel combustion in homes and places of work, but also dissipated from vehicles

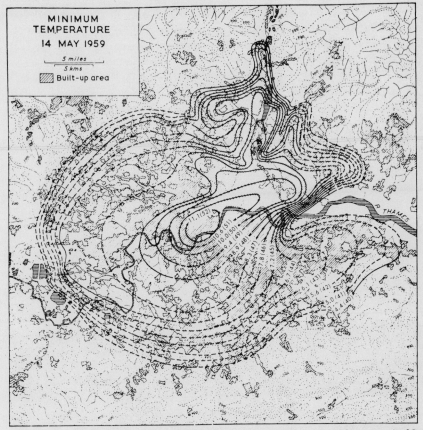

Figure 16. Minimum temperature distribution in London on 14 May 1959, in °C
with °F in brackets. From Chandler (1965).

and emitted from the cooling systems of power stations and from factories
(see SMIC, 1971; FATE, 1974; SCEP, 1972; Energy and the Environment,
1974). These releases augment the effect of the concrete and brick surfaces
in a city, which absorb more heat by day, and radiate it more readily by
night than a vegetated surface does. The result is a 'hot spot' or 'heat island'
over a city both by day and night, but most evident at night under calm air
conditions, especially when there is an atmospheric inversion (Chandler,
1965, Peterson, 1971). Figure 16 illustrates the situation in London in May
1959. It has been calculated that the annual mean temperature in the centre
of a large city like London is now 0.5–1 °C higher than in the adjacent
country (Landsberg, 1962). New York is said to emit seven times as much
heat as it receives from the sun (MacIntyre, 1973). City centres may thus
become more attractive habitats for plant and animal species intolerant of
cold – and the massed flight of starlings to roost in central London has been

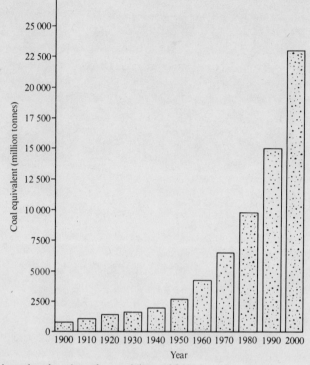

Year

Figure 17. Actual and projected growth in world energy consumption. From United Nations data, used by Ward and Dubos (1972).

suggested as one of the results of snug warmth of the city at night. Thunderstorms can build up over such heat islands, giving massive local rainfall and flash floods, as in Hampstead in 1975 when about 100 mm of rain fell in a few hours (NERC, 1977a).

There has been concern that the continuing growth in energy generation, if present trends (Figure 17) continued, might affect climate over wider tracts of land or even globally. Some authors calculate that it is the finite capacity of the earth to dissipate waste heat into space that sets an outer limit to technology and population, calculated by Fremlin (1964) at 60×10^{15} people (or $120/m^2$). There is an element of the absurd in the more extreme of these extrapolations. At present there is no evidence of a global effect. Even if the whole world reached current American living standards, and supported a population three times the present, the total technological emission of waste heat would only be 0.1 % of that radiated into space by the earth. For the foreseeable future the position will remain one of localised urban heat islands, but these will be liable to expand and coalesce if standards of living, energy consumption and populations continue to rise.

4.3.2 *Heating the waters*

Waste heat is also emitted to lakes, rivers and seas through the discharge of warmed water, used to cool power stations and factories, and the process may also affect flow and salinity. The scale of the phenomenon depends on the type of cooling system used, the size of the plant, and the temperature and dilution capacity of the receiving waters. 'Direct cooling' in which the water is circulated through turbines and then passed straight back into the environment some 8–10 °C hotter than when it came in, is employed only on coastal and lake shore sites in Britain (Hawkins, 1974). Large inland power stations using closed circulation systems with cooling towers can be sited along rivers with relatively low flows and yet raise the temperature of the river at the point of discharge to no more than around 28 °C, which, although a very high temperature by British standards, avoids the creation of zones unfit for life. Ponds in which hot water is cooled to an acceptable temperature prior to release form an alternative system. The river zones warmed by the released water may be altered ecologically, because the rates of most chemical and biological processes depend on temperature (Chapter 5; Lee and Veith, 1971; Nursall and Gallup, 1971). However, these effects need not be adverse (in some areas fish populations may even increase) if siting is careful (Beauchamp, Ross and Whitehouse, 1971). There is, of course, loss of water by evaporation – up to twelve million gallons/day, equivalent to 1% of the summer flow of the River Trent, in a 2000-megawatt power station (Energy and the Environment, 1974). The cascading of water in cooling towers can improve oxygenation, as was demonstrated on the River Trent in the drought of 1976.

4.3.3 *The transmission of radiation through the atmosphere: possible effects of carbon dioxide*

One of the most publicised possible effects of environmental pollution is the change in global temperature which might be caused by the continued accumulation of carbon dioxide or increased levels of fine particles and water vapour in the air. A World Meteorological Organisation working group (WMO, 1975), the Study of Man's Impact of Climate (SMIC, 1971), the Royal Commission on Environmental Pollution (1971), SCEP (1970) and SCOPE (1977) all consider the 'greenhouse' effect of carbon dioxide as the most plausible of all potential global effects of pollution.

Carbon dioxide is a final oxidation product of fuel combustion. It is a naturally abundant compound, intimately involved in biological cycles, and essential to the maintenance of life (Chapter 1). At the beginning of the industrial era it was present in the air at sea level at a concentration of about 265–270 ppm. In 1976 the concentration was about 325 ppm. Regular

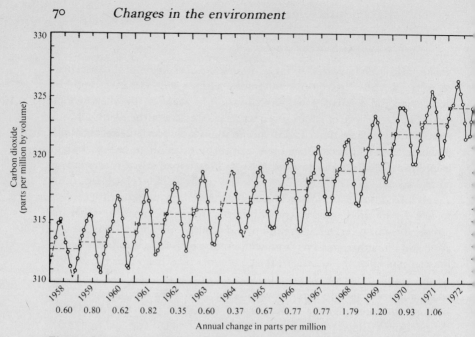

Figure 18. Increase in carbon dioxide concentrations at Mauna Loa Observatory. Source: Department of Commerce, National Oceanic and Atmospheric Agency, based on data provided by C. D. Keeling, Scripps Institute of Oceanography, sponsored by National Science Foundation.

measurements made since 1958 on the summit of Mauna Loa, Hawaii, and more recently also at the South Pole, confirm its continuing upward trend and incidentally, because they agree so well, its rapid world-wide mixing. Figure 18 shows this trend at Hawaii. The oscillation is due to uptake by plants in the summer 'growing season'. The net annual increase between 1958 and 1968 was about 0.7 ppm, but the rate seemed to be accelerating towards the end of this period and by 1971 had reached about 1 ppm/year: at the South Pole this acceleration had not manifest itself by 1971, and the annual increment remained about 0.7 ppm. Since 1971 there has been some indication of a reduction in the annual increment in the northern hemisphere.

There is an anomaly about this trend. Calculations suggest that if all the carbon dioxide released from the burning of fossil fuels had remained in the air, the annual increase would have been about twice that observed. Where is the 'missing half'? Almost certainly it is in the sea. Carbon dioxide dissolves and diffuses readily in water. The total amount of carbon dioxide in the air is only about one-sixtieth that in the ocean. In the sea there are complex interchanges between dissolved carbon dioxide and a range of inorganic carbonates and bicarbonates, and while rapid mixing is only a feature of the upper 100 m or so of ocean, there is a slow interchange with the deeper waters, and marine sediments act as a 'sink' for carbon dioxide. The earth's abundant limestone and chalk rocks include massed skeletons of microscopic marine life and witness how important the sea has been in this way in the past. Figure 19 (from SCOPE, 1977) indicates the predominant role of sediments as stores of carbon, containing a thousandfold more than do fossil fuel reserves. It also indicates that photosynthesis on land and at sea takes up annually about a quarter of the carbon in atmospheric carbon dioxide (150×10^9 tonnes out of 670×10^9 tonnes). There is much more carbon in living organic matter on land than in the sea, but the sea is the main inorganic sink. However, uptake by solution and photosynthesis are almost balanced by the returns through respiration, decomposition and losses from solution. The input from fossil fuel combustion (3.6×10^9 tonnes/year), seen in this context, is tiny compared with the other fluxes and the stores of carbon.

We still lack a complete model for these interactions, and so cannot calculate how far further emissions of carbon dioxide at a higher rate would outstrip the capability of the sea or photosynthesis to take up the surplus. The position is further complicated by man's impact on ecological systems through forest clearance and soil oxidation on the one hand and reafforestation on the other. The best estimates, assuming that the sea will continue to take up 50–60% of the increment of carbon dioxide added to the air from fossil fuel burning, suggest that for a time atmospheric concentrations will continue to rise and some models suggest a fourfold or even eightfold increase over the next 150 years (WMO, 1975; SCOPE, 1977). But on a geological time scale man's impact on carbon dioxide levels in the biosphere as a whole appears negligible when set against the amounts in circulation between air, ocean and sediments (Junge, 1972).

In the short term (which in this context means decades or centuries), however, it is likely that the fraction of the emitted carbon dioxide that remains in the air will affect world temperatures. This is because the carbon dioxide in air has a 'greenhouse effect': it lets through incoming radiation from the sun but reduces the radiation of energy back into space. Machta (1972) has computed that if present trends continue, atmospheric levels of carbon dioxide would reach 380 ppm by 2000 and an extension of the analysis by WMO suggests a level of 450 ppm by 2050. Some authors have predicted steeper rises to higher levels. Manabe and Wetherald (1975) have calculated that such changes might raise mean temperatures in the lower atmosphere by 0.5 °C in 2000. This would be enough to affect agriculture and natural ecosystems (especially in marginal areas) and also influence energy consumption in buildings, although the change is well within the range of natural climatic oscillation and might indeed be cancelled out by the cooling of world climate that many authors consider the present

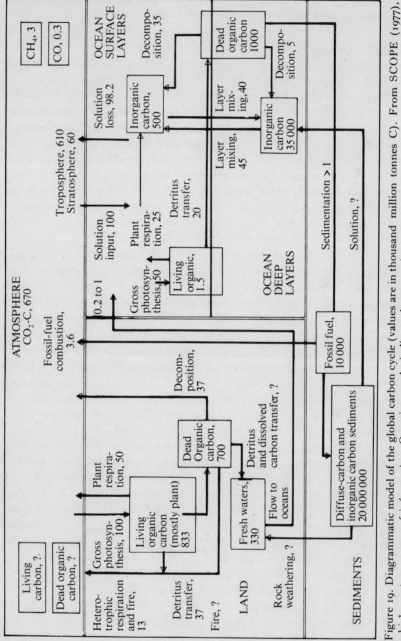

Figure 19. Diagrammatic model of the global carbon cycle (values are in thousand million tonnes C). From SCOPE (1977), which quotes sources of information. Question marks indicate that no estimates are available.

natural trend (e.g. Weiss and Lamb, 1970; Lamb, 1977). Later figures (WMO, 1975; SCOPE, 1977) imply that a doubling of atmospheric carbon dioxide would raise mean global temperatures by 1.5–2.5°C, but the relationship is not a linear one and successive doublings might have a diminishing impact. Such changes in temperature would obviously have a profound effect and it is consequently important that research into these problems continues.

There are factors which may operate to slow the process. The present rapid rate of forest clearance, using primitive methods (including fire), may not be sustained for many more decades and may in part be countered by management of surviving woodlands for higher biological productivity. More significantly, although developed countries are at present heavily dependent on fossil fuel and many, including the United Kingdom, have substantial coal reserves, this situation will not last for ever. The proportions of nuclear, wave, solar and other energy sources that will be used in the future are uncertain, but the result of the ultimate switch from fossil fuels will undoubtedly be a slowing of carbon dioxide release, while the operation of the natural sinks in the ocean should act to stabilise and probably then lower the levels in the atmosphere. It seems unlikely that we shall go on releasing carbon dioxide until we cook ourselves, or flood ourselves by melting low-lying polar ice masses.

There are other possible brakes on the warming that might result from rising carbon dioxide. One of the most likely is that any increase in temperature might increase cloudiness, and this would certainly have a major effect on radiation balance. A 2% increase in mean cloud cover over the world could reduce mean temperatures as much as they would be raised by a doubling in carbon dioxide. Man already injects dust and vapour into the upper air. In the early 1960s the Mauna Loa records seemed to show a rise in particulate levels high up, and it was thought that these might balance any warming due to carbon dioxide by screening out sunlight. However, the amount of dust in the upper air fell in the early 1970s and it is now thought that the Hawaiian records were being affected by the 1963 eruption of Mount Agung in Bali. Elsewhere in the northern hemisphere, there is some evidence that particulate levels in air and high clouds produced by aircraft vapour trails have increased, and in parts of the USSR, the Swiss Alps and North America there has been a decline of up to 10% in the amount of radiation reaching the ground. This effect appears, however, to be confined to the northern industrial belt and may be of little significance on a global scale.

4.3.4 *Transmission of radiation: possible effects of aircraft, chlorofluorocarbons and oxides of nitrogen*

Another, quite distinct, possible effect on radiation has been alleged to be likely to result from an increase in the number of aircraft flying in the lower stratosphere. This has been a cause of recent controversy over the environmental acceptability of supersonic transports (SSTs). In order to achieve high speeds, skin friction must be minimised and this means flying in the stratosphere, at altitudes hitherto invaded only by small numbers of small-sized military aircraft and by passing rockets. SSTs are not only bigger, but programmed to spend many hours a day in the stratosphere, while their higher speed naturally demands more energy and hence more fuel consumption. Should they become the normal long-range transport craft, flying in hundreds, they would certainly augment emissions to this zone, which (as Chapter 3 indicated) does not mix rapidly with lower layers of air. Pollutants injected there, like the chlorofluorocarbons, tend to stay for a long time.

The problem arises because while such aircraft continue to be impelled by hydrocarbon fuels (for which there are no obvious immediate substitutes), their exhaust emissions will inevitably contain oxides of nitrogen and unburned hydrocarbons. As with motor car exhaust at ground level, oxides of nitrogen and hydrocarbons are involved in photochemical reactions which also involve ozone, and some calculate that a threat might be posed to the ozone layers in the stratosphere which screen out ultra-violet radiation from the sun. If the screen were weakened and more ultra-violet got through, there could be an increase in skin cancer. The problem was reviewed by a Climatic Impact Committee of the United States National Academy of Sciences (CIC, 1975) which suggested that allowing 100 current-generation supersonic aircraft (far more than at present planned) to fly continuously in the stratosphere could raise the incidence of skin cancers by about 1½ % – this is by a few tens of cases per million of population, a small proportion of whom might die. A good deal of research is now going on to find out whether these fears are in any way justified (CIAP, 1974a, b; Goldsmith, 1974; DOE, 1976a; COMESA, 1975). There has been controversy over the mathematical models used – and this is inevitable because we are still ignorant of the details of many of the interactions involved. The general conclusion is, however, that no significant effect is likely from the fleet of aircraft likely to fly in the lower stratosphere in the forseeable future. Even the levels of increased radiation predicted from 100 aircraft would be no more than an individual incurs when moving from a district as cloudy as Glasgow to the relatively higher sunshine of southern England, and considerably less than voluntarily accepted by enthusiastic sunbathers. Most published analyses suggest that

75   4.3 *Physical effects*

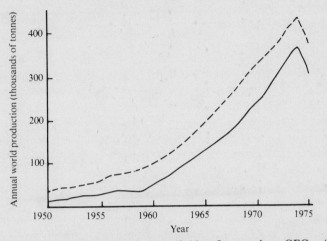

Figure 20. Estimated annual world production of chlorofluorocarbons CFC 11 (solid line) CFC 12 (broken line). From DOE (1976a).

the impact on stratospheric ozone concentrations likely to result from chlorofluorocarbons is likely to be more signficant.

In 1974 about 800 000 tonnes of chlorofluorocarbons were manufactured in the world (Figure 20): by 1975 the figure had dropped to a little under 700 000 tonnes. Since 1931 about 8 000 000 tonnes have probably been made. They are released to the air in the boundary layer when aerosols are used or refrigerator cooling systems leak. Since 1970, concentrations in the troposphere have increased by a factor of 1.5 to 2 (Figure 21). Because of their slow breakdown (Chapter 3) the bulk of all the CFCs so far emitted will still be present in the atmosphere, and even immediate restrictions would not be followed by significant reductions for some years. These chlorofluorocarbons accumulated in the stratosphere may also increase atmospheric absorption and emission of infra-red radiation, retarding heat loss from the earth and hence acting like carbon dioxide to produce climatic warming, but the scale of the phenomenon is uncertain. The best recent estimate (CISC, 1976) is that continued release of these compounds at the rate that prevailed in 1973 could by 2000 produce about half the effect on climate likely to be caused by carbon dioxide.

The third hypothetical modification of stratospheric ozone concentrations could be caused by oxides of nitrogen, whose natural levels (produced by lightning and biological decomposition) are augmented by fuel combustion and the manufacture and use of nitrogenous fertilisers. One unpublished calculation by M. B. McElroy (referred to in DOE, 1976a) suggested a possible 30% reduction in ozone concentrations as a result of predicted increases in world fertiliser production, although such assertions have been questioned. Machta (1976) attempted to calculate the

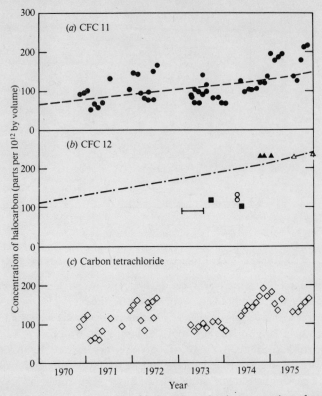

Figure 21. Measured concentrations of halocarbons in the troposphere from 1970 to 1975. The various symbols in (*b*) represent the results of different authors. From DOE (1976a).

dose and harm commitment that might result from ozone depletion due variously to supersonic and subsonic aircraft flying at high altitudes, fertilisers and chlorofluorocarbon aerosols and refrigerants, and his results are reproduced in Table 10. Clearly the range of variation between alternative estimates is very great, reflecting the uncertainties in the whole process of calculation, and (as Chapter 5 indicates) the 'harm commitment' figures assume no adjustment to the present pattern of exposure to incoming radiation which is very largely affected by the voluntary behaviour of individuals, for example through sunbathing.

Despite these doubts, it remains true that emissions of carbon dioxide, oxides of nitrogen, chlorofluorocarbons and particulates to the air could conceivably cause global and regional changes in climate and in the penetration of potentially harmful radiation, and they are the only types of pollution for which such widespread effects can at present be postulated reasonably. Because of the all-embracing nature of the possible climatic

Table 10. *Estimated environmental effects (dose commitment) and incidence of and mortality from skin cancers (harm commitment) due to steady state release of selected chemicals*[a]

Chemical	Input (g/year)	Altitude of input (km)	Source	Dose commitment: % Ozone depletion	Dose commitment: % Increase UV-B	Harm commitment (additional cases per year)[b] Incidence: Non-melanoma skin cancers	Incidence: Melanoma skin cancers	Mortality: Melanoma skin cancers
NO_x	156×10^{11}	10.5	400 wide-bodied subsonic aircraft	0.082	0.12	2000	40	10
NO_x	1×10^{11}	16–18	100 Concorde/TU-144 aircraft	0.11–0.57	0.17–0.86	2500–15200	50–250	10–50
NO_x	164×10^{11}	19.5	100 large supersonic aircraft	3.27	4.91	75000	1500	300
N_2O	$2 \times 10^{14}(N_2)$	0	Fertilisers in A.D. 2000 and constant thereafter	< 1.8–23	< 2.7–35	< 40000–500000	< 900–10000	< 150–2000
F[c]-11, F-12	2×10^{11}, 3×10^{11}	0	Aerosols, refrigerants, etc. at 1973 rate	6.5–18	9.8–27	150000–400000	30000	600–1500

From Machta (1976).

[a] Estimates assume a constant world light-skinned population of 10^9 persons.

[b] Based on current non-melanoma skin cancer incidence of 150/100000 light-skinned persons per year; melanoma incidence of 4/100000; melanoma mortality of 1.5/100000.

[c] F-11 and F-12 are chlorofluorocarbons.

consequences, altering habitats and biological productivity all over the planet, such impacts are naturally regarded with concern. Even if slight in terms of mean temperature change, local effects on rainfall, cloud and 'growing season' could be profound. They could cause shifts in the boundaries of deserts and the limits of cultivation about the poles. Substantial warming could affect the balance of polar ice caps, and while the great ice plateaux of Antarctica and Greenland are very cold because they are very high, melting at low level could have a measurable effect and even a rise in sea level of a metre or two would be costly and inconvenient. 'Teleconnections' between the climate in widely separated places are well known (SCOPE, 1977; Lamb, 1977). Changes of sea temperature in parts of the North Pacific are, for example, apparently associated with fluctuations in European weather, and much larger changes such as the disappearance of sea ice in broad areas of the Arctic basin would probably have major repercussions.

In the long term, all the changes so far postulated seem to lie within the range to which organisms and ecosystems have been exposed during the process of evolution, so that adjustment rather than catastrophe is the likely outcome. But some of the social implications even of minor and local change cannot be ignored, in a variable world with some regions poised precariously between fertility and drought, and under circumstances where the main beneficiaries of the pollution-generating activities and the main victims of any damage they cause are likely to live in different places. This theme of international responsibility for pollution and its consequences is returned to in Chapter 9.

4.4 Chemical changes in the air: trends in other atmospheric contaminants

Chlorofluorocarbons and oxides of nitrogen react so as to modify the concentration of a normal constituent of the atmosphere, with consequent effects on radiation transmission. They exemplify substances for which the medium is, in effect, a 'target'. In contrast, many other pollutants whose levels in air or water are raised by man exert little effect on the medium and cause concern chiefly because they influence targets exposed within it.

4.4.1 Sulphur dioxide and smoke

Sulphur dioxide overlaps the two categories. It can react, for example, with ammonia to form an ammonium sulphate haze which is not only corrosive but obscures sunlight, and this haze was formerly a notorious feature of some areas with large chemical industries (e.g. 'Teesside mist'). It also affects targets when deposited directly or in solution in rain.

There is no evidence that sulphur dioxide accumulates in the atmosphere (and much that it does not). Its residence time there is only a matter of days. Raised emissions are balanced by raised depositions. But emissions have been rising and are predicted to go on doing so as long as mounting energy demands are met largely by burning fossil fuels in the traditional way. Some forecasts (Holdgate and Reed, 1973) indicate that global emissions will nearly double by the end of this century, most of the increase being in the industrialised zone of the northern hemisphere which is already the most polluted (Table 11). Influences working against this trend include a diversion to other energy sources, the increasing use in this region of Alaskan and North Sea oils, both lower in sulphur than many African and Middle East crudes, and growing pressure for desulphurisation of fuel or of emissions. The available techniques include fluidised bed

Table 11. *Atmospheric sulphur budget*

	World output or removal (million tonnes S per annum)
Sources	
1. Natural sources	
(a) biological decomposition (land and sea)	32
(b) sea spray	44
(c) volcanoes	3
2. Man's activities	65
(in NW Europe alone)	(13)
	144
Sinks	
1. Dry deposition (land and sea)	
(a) 'natural' sulphur	6
(b) 'man-made' sulphur	32
2. Wet deposition (land and sea)	
(a) 'natural' sulphur	73
(b) 'man-made' sulphur	33
	144

From SCOPE (1977).

combustion in power stations and the scrubbing or catalytic treatment of flue gases (Beer and Hedley, 1973). The fluidised bed technique may be cheapest, when installed at the outset in new plant, and has the added attraction that it may improve combustion efficiency by some 10%. The obstacle to its wider adoption is the still unproven state of the technology and the capital cost of retrospective conversion of existing equipment. Of the methods for desulphurising waste gases, catalytic systems (which have been used fairly extensively in the United States) appear more expensive than those involving the addition of finely dispersed absorbents like limestone and dolomite, or wet scrubbing. All, however, have certain drawbacks including the disposal of sulphur, absorbent and water after treatment.

In Britain, sulphur dioxide and smoke together have long been regarded as the most significant air pollution hazard (Royal College of Physicians, 1970; Royal Commission, 1971; Ashby, 1972; DOE, 1972a). British coals have a relatively high sulphur content, and the problem was aggravated when combustion took place in old-fashioned domestic grates with poor thermal efficiency. It was the irritant effects on the lungs of domestic smoke and sulphur dioxide, held in a moist fog below an inversion, that caused the great smog of London in 1952–3 when some 4000 excess deaths occurred, especially among the elderly and those with cardiac and

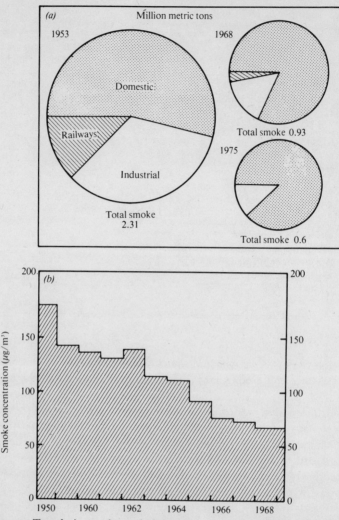

Figure 22. Trends in smoke emission and concentration in Britain. (*a*) Smoke emissions in the UK for 1953 and 1968 with a prediction for 1975. (*b*) Average smoke concentration near ground level in the UK, 1958, 1968. From Royal Commission (1971).

respiratory disorders. Since the report of the Beaver Committee (Command 9322, 1954) investigating that event, and the Clean Air Acts of 1954 and 1962, local authorities have declared smoke control areas in the worst urban zones. Linked to the increased availability of socially preferred cleaner fuels like oil and gas, to urban renewal and more efficient central-heating systems, these policies have led to a fall in total smoke

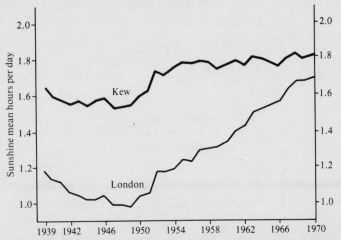

Figure 23. Trend of winter sunshine (December to February) at London Weather Centre and Kew Observatory. After Royal Commission (1971).

emissions in Britain from some 2.7 million tonnes in 1960 to 0.77 million tonnes in 1970 (Figure 22). There was a minor smoky smog incident in London in 1962 (Royal College of Physicians, 1970), but since then no winter has produced conditions reflected in substantially increased hospital admissions or deaths. Over 69 % of all houses in Great Britain in areas identified as 'black' in 1956, and 90 % of all houses in Greater London are now covered by such smoke control orders. The effect on winter sunshine in the centres of cities like London and Manchester is well known (Figure 23). In Glasgow such sunshine has increased by 60 % in the last 20 years, and today, when we clean our civic buildings they stay reasonably clean. More importantly, smoke does not appear to be killing people any more in London. But there and elsewhere there are urban 'black areas' where smoke levels are undesirably high. The WHO short-term objective (see Chapter 6) of a maximal 250 μg smoke/m^3 in the presence of 500 μg SO$_2$/m^3 (as a 24 hour mean) is still exceeded on a number of days. The WHO long-term objective was to attain annual means so low that even sensitive plants are unharmed. On this basis we have a long way to go in Britain although continued effort is being directed towards reducing the size and severity of the 'black spots' and this is being pressed at local level in some cities. Recently, 'guidelines' of 60 μg SO$_2$/m^3 as an annual mean and 40 μg/m^3 for particulates have been promulgated by the Greater London Council: these are close to the WHO long-term objectives, but are likely to take some time to attain.

The total emissions of sulphur dioxide in Britain rose in the years up to 1970, to a peak of 5.5–6 million tonnes/year (Warren Spring, 1972).

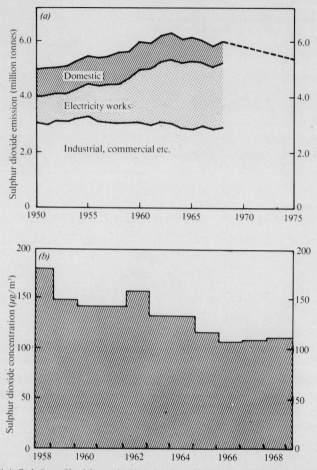

Figure 24. (*a*) Sulphur dioxide emissions in the UK, 1950–68, with prediction to 1975. (*b*) Average sulphur dioxide concentrations near ground level in the UK, 1958–68. From Royal Commission (1971).

However, changes in domestic fuels and the increasingly central generation of power has meant that more and more of this sulphur has been put out of tall chimneys at large plants. Concentrations near the ground in most central urban zones have therefore fallen by about 30% over the last 20 years (Figure 24). They still remain above the desirable 500 $\mu g/m^3$ in central London and other large towns for up to 20 days/year. The Greater London Council's stated objectives for eliminating this residual 'hot spot' in London have already been mentioned: in addition, the City of London has prohibited the burning there of fuel containing more than 1% sulphur. This is unlikely to have any great effect because of the small extent of the 'square mile', although it should lower levels on still days when locally emitted gases tend to be trapped in the man-made canyons of the streets. But it provides a precedent (paralleling similar action in Stockholm, Rotterdam and Paris) and could be followed in comparable urban centres.

The sulphur dioxide story is a good example of scale variables. Initially the worst problems detected were local – smoky irritant smogs in towns. Here, trends in sulphur dioxide levels are now downwards. Today we are concerned with regional or national problems of impaired plant growth or fresh-water fisheries over part of a continent – and at global level, we have already noted that the trend in emissions in upwards. There remains a possibility that while winning the local, intense, urban pollution battle we may be losing the wider, less intense ecological one. Since controls on sulphur levels in fuels and emissions are already being mooted and the technology to abate emissions exists, the upward trend of emissions in industrial zones could be reversed if the benefits prove to outweigh the costs.

4.4.2 *Gaseous emissions from vehicles*

Emissions from transport are another major source of air pollution through fossil fuel combustion (Royal Commission, 1971; Bertodo, 1974). In Britain, black smoke from badly maintained diesel engines came under regulation first because it is unsightly, a contributory cause of road accidents, and smells. There has been more controversy over the scale of hazard posed by emissions from petrol-driven cars: carbon monoxide, unburned hydrocarbons, oxides of nitrogen and the lead added to petrol as 'anti-knock'.

Carbon monoxide, like sulphur dioxide, is not accumulating in the air (although some may well be oxidised there to carbon dioxide and so swell the upward curve of that gas, while more may be reduced by bacteria to methane). Moreover, man's contribution to emissions appears to be considerably less than that of natural sources. Holdgate and Reed (1973) quoted estimates of 5×10^9 to 10×10^9 tonnes/year produced naturally in contrast to 0.2×10^9 to 0.02×10^9 tonnes produced by man (IRCCOPR, 1974, estimates 3×10^9 and 0.6×10^9 tonnes/year respectively). There seems no cause for concern at global level about this gas, while locally the best evidence (e.g. Lawther, 1973; Ashby, 1973; Lawther, 1975) is that it is a hazard to healthy people only in confined spaces (like a tunnel or garage forecourt): in a busy street the level it attains in the blood (1–3% of saturation) is commonly less than that produced in a moderate cigarette smoker (heavy smokers can attain up to 5–10% saturation) and the effect it has in blocking the part of oxygen carrying capacity of the blood is fully reversible. But 3–4% saturation could be a hazard to someone with cardiac weakness, just as a marginal increase in smoke and irritant sulphur dioxide can weigh the scales against a bronchitic or a sufferer from emphysema.

This, indeed, is the critical problem and it is a matter of risk assessment leading to a particular standard (as Chapter 6 explains). At the margin, someone near the point of cardiac or respiratory collapse can undoubtedly be tipped over into death by a minor impairment in the oxygen capacity of the blood or the oxygen uptake facility of the lung. But other forms of stress can also tip the balance – undue exertion and discomfort in the jostle of public transport, a bitter spell of weather, a period of poor diet or too many cigarettes. We cannot expect that pollution will never shorten life marginally. The aim is to ensure that the number of cases is small, the shortening slight, and the scale of the hazard known.

The other products of motor cars are potentially more troublesome. Hydrocarbons and oxides of nitrogen interact to produce oxidants like ozone and in California, in the bright sun and stable air around Los Angeles, the devastation such oxidants can cause to sensitive vegetation is well known (Figure 25). For some years a comfortable myth that this 'Los Angeles smog' only occurred in Los Angeles has prevailed. Research shows that the facts are otherwise. In the USA similar oxidant levels are typical of major cities, especially under similar climatic conditions, while enhanced oxidant levels have been recorded over whole states, such as Pennsylvania. In Europe, high oxidant levels have been found in Rotterdam and can be predicted in Rome, Turin, Milan, Paris and Munich. Britain has a less sunny and more turbulent climate – but summer sunshine in London compares well with autumnal and spring levels in Los Angeles, when smog is frequent (Atkins, Cox and Eggleton, 1972). Recent monitoring has shown that the US air quality standard (that an hourly mean ozone level of 8 parts per hundred million should not be exceeded more than once a year) is broken in the West End of London (Derwent and Stewart, 1973). In 1972–3 levels passed 8 pphm on 22 days, for a total of 69 hours, and passed 10 pphm for 23 hours. Even in the rural setting of Harwell, remarkably high ozone levels have been discovered, especially with anticyclonic conditions and a southerly drift of air, perhaps bringing oxidants from mainland Europe. There is insufficient monitoring to delimit the problem in Britain, but clearly the earlier period of confident reliance on the protective influence of our climate is past.

4.4.3 *Emissions of metals, droplets and particles*

Lead, oil and pesticides all exemplify the wide dispersal of solid particles or droplets of pollutants by air. The wide dispersion of the smallest fraction of lead particles has been discussed already in Section 3.2.1. The fact of this dispersion is unchallenged, and so is the clear evidence of an

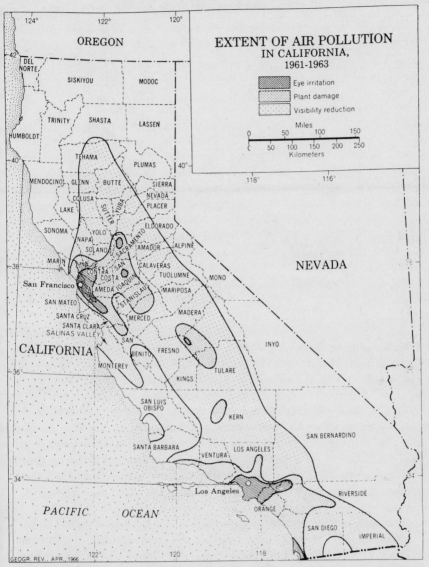

Figure 25. Extent of air pollution damage in California. From *Man's impact on environment* by Detwyler. © McGraw-Hill Book Company. Used with permission of McGraw-Hill Book Company.

overall upward trend in lead emission and deposition. Fresh samples of moss from Sweden have been compared with others dating from the early years of the Industrial Revolution and the modern samples contain about 4 to 7 times as much lead as the early ones.

Even more compelling evidence comes from snow samples from

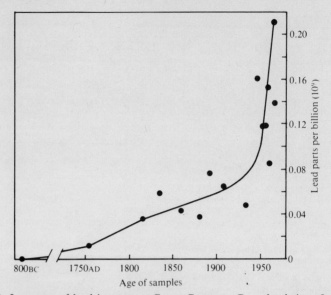

Figure 26. Increases of lead in snow at Camp Century, Greenland since 800 BC. After Murozumi *et al.* (1969).

Greenland (Figure 26: Murazumi *et al.*, 1969). The earlier part of the upward trend clearly coincides with the industrial Revolution. The rapid rise since 1930 coincides with the introduction of lead compounds as additives to petrol. Similar rises are now showing themselves in snow cores from the Antarctic, but concentrations there are lower, and both these features probably result from a barrier to atmospheric transport between the two hemispheres, whose tropospheric circulations do not mix very much (Chapter 3). Thus the trend in atmospheric transport of lead in recent decades is probably more marked in the northern hemisphere than globally, as are the trends in the emission and deposition of sulphur dioxide.

The damage to vegetation and unpleasant effects on man caused by the 'oxidant smog' derived from vehicle emissions has led to stringent controls in the United States (CEQ, 1971). In Europe those adopted have been less rigorous, but sufficient to ensure that the trend in all pollutants emitted by vehicles no longer continues to rise in proportion to the increasing number of cars on the road. The first European measures were the adoption of a crank-case re-circulating device to reduce hydrocarbon emissions, and a limit on carbon monoxide. The maximum permitted level of lead in petrol is also being reduced: in Britain the target is 0.45 g/l in 1978 and the achievement of a common European Community standard of 0.40 by 1981 (DOE, 1976e).

Rather similar considerations apply to DDT, DDE (its derivative) and other persistent organochlorine pesticides like dieldrin, aldrin or benzene hexachloride. That these substances are widely distributed has been highlighted by the detection of residues in the fat of penguins, seals and other organisms in the Antarctic. Some of the residues may have entered the Antarctic or sub-Antarctic system because seabirds that breed there range widely at other seasons and some, like the Arctic tern, great shearwater, and New Zealand dusky shearwater even enter northern hemisphere waters. The discovery of residues of DDT in the eggs of a small land bird (the 'starchy' *Nesocichla eremita*) confined to the Tristan da Cunha islands may be accounted for by its habit of eating seabird eggs, and by the fact that great shearwaters, which breed in numbers in the Tristan group, winter in the North Atlantic. Seaborne contamination and local releases from manned stations may have played a further part, but the most likely pathway of these pesticides to the far south is as airborne droplets. Like lead, organochlorine pesticides are being increasingly limited in northern industrial countries, where usage is now falling fast – primarily because of concern to avoid the unwelcome biological side-effects they have caused among wildlife in those countries. However, DDT is still of great importance in malarial control in many tropical countries, and while its use is likely not to increase as dramatically in coming years as it has since it was generally introduced in the early 1940s, because of its cheapness and effectiveness and the lack of good substitutes, it is likely to continue to be applied for many years in tropical parts of the developing world. Its persistence in the biosphere (with a half life of around 30 years) means that it will probably be many decades before the global trend in residue levels is downwards. But it must be emphasised that the harmful effects of organochlorines are, like those of lead, localised in areas of fairly high concentration. The 'turn round point' is likely to be well below the threshold of general ecological damage.

No attempt will be made here to describe the concentrations of, and trends in, the vast number of substances emitted to air by various industries (Chapter 2). Most of these pose local problems only, and are the subject of increasing controls (Chapter 8 and HSE, 1975). Under these controls, many notorious local disamenities have been brought to an end. For example, 'Teesside mist' has largely disappeared following a reduction in ammonia leaks. In the 14 years from 1958 to 1974 power stations in Britain burned 35% more coal but emitted 82% less grit and dust, and cement production rose by 60% but total emissions were reduced by 64%. This does not mean that we have reached perfection. There are still local 'hot spots' where contamination is undesirably high. For example, a survey in 1972–3 (DOE, 1974b) showed that around urban works using lead, especially in dusty form, lead levels in the streets and gutter dusts could reach as high as 5–20% and blood levels were raised not only in lead workers and their families but in people living within several hundred metres of the works. Noise and smell remain a nuisance around many works. Carbon black particles deposit around others. Fine dust is a problem even with a modern cement works using 'best practicable means'. These residual problems arise partly because abatement technology is still not fully efficient. Partly they result from the sheer size of many modern works. Electrostatic precipitators can be designed to give 99.5% efficiency in stopping fine dust, but such devices do not give their full design performance throughout the year, and even when they do, the residual 0.5% can represent a large tonnage of dust, much of which falls in the immediate neighbourhood.

4.5 Changes in rivers, lakes and estuaries

Changes in the proportions and concentrations of dissolved and suspended substances in fresh and estuarine waters are more widespread than the thermal changes already described. Like air pollution, many of these changes are special features of densely populated industrial areas, but there are some regional effects, for example on great river systems like the Rhine, Danube or Rio de la Plata or in the North American Great Lakes. The latter show degrees of deoxygenation that can clearly be related to the extent of pollution. In 1960 the least affected, Lakes Superior and Huron, remained nearly saturated with oxygen: Lake Ontario was only 50–60% saturated at the deeper levels in winter, and Lake Erie 10–50% saturated. In late summer about 2600 square miles of the bed of Lake Erie were deoxygenated (Beeton, 1965).

The trends in British fresh and brackish waters, while lacking such extensive changes, provide a fairly typical national example.

4.5.1 'Eutrophication' and sewage

There is no doubt that the lakes, rivers and estuaries of Britain have changed very greatly since their natural state at the end of the last glacial period. Some of the changes were natural and inevitable. For example, glaciated valleys like many in Scotland, Cumbria and Wales, contain lakes which have very little dissolved salts. These lakes fill basins scooped from the rock by ice, or impounded by crushed rock and moraine. Rainfall is high, and the water well mixed in spring and autumn: the result is a rapid turnover of the lake's contents. Such 'oligotrophic' lakes contain little more than distilled water. But the trend over the centuries is for any lake to silt up. Sediment is deposited – some of it peaty, organic matter. Reeds push out from the shore. Organic matter increases and levels of nutrient salt also rise. Nitrogenous and phosphatic matter is contributed in drainage from the lake basin (some of it derived from animal waste matter, and some from decomposition). Unless these inputs are removed in outflow or sediments, the trend is towards richer nutrient status – towards 'eutrophication'. In Chapter 1 it was pointed out that man, through the discharge of sewage and through increasing the run-off of nitrate and phosphate from the land (through forest clearance, intensive husbandry and the addition of fertilisers) accelerates this process and may cause an over-shoot beyond anything in nature.

The upland lakes of Britain have escaped this process, although many have been contaminated with metallic salts derived from early mining. Today, however, their waters and that of the streams draining them remain generally pure, which (coupled with the high rainfall) is one reason why so many reservoirs are placed in the hills despite the long aquifers needed to convey the water to the cities. But the lowland lakes and rivers have suffered much more from nutrient enrichment, with the consequent over-growth of plant matter, increased production of dead material, and deoxygenation. The estuaries are also affected by this process, and many (like Poole harbour) show bright green fringes of alga around high water and floating debris in the sluggish tides.

A century ago sewage treatment was virtually unknown. In primitive towns and villages, faeces were transported by bucket to the fields, as a useful fertiliser, or left as dunghills in the streets, to be carted away periodically to a dump from which it no doubt often found its way into drainage systems. The development of guttering that could be flushed by water from a conduit (as in eighteenth-century Cambridge) or later of drains that carried rainwater under the streets, bearing muck with it, naturally conveyed the wastes to river systems and the addition of water-borne sewerage accentuated the trend. Hence in the mid to late nineteenth century many rivers and canals had lost their earlier richness in fish (emphasised by the legendary protests of apprentices in London and Manchester against being fed salmon every day) and had become stinking and lifeless. In the 1860s it was a sport to light the methane rising from Yorkshire canals and rivers and see blue flames run along the water enveloping the barges as they went – and no doubt drawing language of equal blueness from the afflicted bargees (Pentelow, 1959). The Thames in London could not decompose the sewage shed into its oxygen-free waters, so that faecal matter floated freely past the Parliament buildings and the smell is said to have occasionally demanded the adjournment of proceedings. Some contrast with the time when Henry II's polar bear at the Tower is alleged to have been loosed to fish for itself in the river!

To this organic matter, industry added an increasing volume and diversity of toxic materials, as well as some organic matter which like sewage acted mainly by fertilising the waters to the point where excess growth caused deoxygenation. Brewery and distillery wastes and slaughterhouse and leatherworks effluents come into this category while the wastes of metal-working, ceramics and chemical plants as well as mining can be directly toxic. By the end of the century the state of the inland waters and estuaries in the industrial zones, with their fast-mounting populations, was causing serious concern.

The first response was to set up Royal Commissions. Two, on sewage disposal, sat in the 1880s and up to 1910. They recommended treatment of sewage prior to release, and that the final effluent which was, moreover, to be diluted tenfold by the river flow, should not have a biological oxygen demand greater than 20 ppm or a similar loading of suspended solids. These standards would, if attained, have undoubtedly dealt with the sewage problem effectively – but they were so stringent by comparison with the actual conditions of the time, and demanded resources on a scale that administrations found it so difficult to provide, that even now they are not fully met by a number of sewage works.

It is, however, true that untreated sewage is now hardly ever discharged into British rivers (although a good deal goes into the sea after only the primary treatment of maceration and screening) (DOE, 1973). The chief exception is in times of storm, in older towns where the storm drains and sewers are linked and when the volume of combined inflow is too much for the sewage works to handle. Under these conditions – fortunately when river flows are likely to be high and turbulence aids oxidation – the storm water with its added load of untreated sewage may be released directly to the rivers. New sewage works segregate the two types of drainage, thereby removing the problem.

Secondary sewage treatment is a method whereby the organic material is broken down by bacterial action in large tanks, the process being speeded up by various methods of aeration, inoculation with active micro-organisms, and controlled circulation. The end product is a liquid lacking dangerous bacteria and also lacking organic matter. But it still has the capacity to cause pollution, for it is rich in nitrates and phosphate and therefore able to enrich aquatic ecosystems. If phosphate loading in particular is to be controlled, as in places like the Norfolk Broads whose sluggish rivers and shallow lakes readily respond with the production of algal blooms, it may be necessary to remove it from the effluent through a tertiary treatment such as flocculation with aluminium hydroxide, precipitating the phosphate as aluminium phosphate. Such treatment is not common in Europe or North America, but may be increasingly familiar in future.

4.5.2 *Overall trends in fresh-water quality*

Extensive surveys of river quality were conducted in Britain in 1958 and 1970 (DOE, 1971, 1972c). Four categories were defined:

Class I   Unpolluted and recovered from pollution.
Class II   Of doubtful quality, needing improvement.
Class III   Poor quality, requiring improvement as a matter of some urgency.
Class IV   Grossly polluted.

The criteria for definition included biological characteristics, especially the diversity of the flora and fauna and the presence of species noted for their exacting oxygen demands. Some of these have been further elaborated in biological indices like the Trent Biological Index, used in monitoring exercises and mentioned in Chapter 7.

The comparison of the 1958 and 1970 surveys confirms a considerable improvement in river quality over that period, while the 1958 situation was far advanced from the murky days at the turn of the century. In 1958 there were 1278 miles of Class IV rivers in England and Wales: in 1970 the mileage had dropped to 952 miles and by 1972 to 832 – only 3.7% of the total mileage in the two countries. At the other end of the scale, the unpolluted mileage totalled 14603 in 1958, 17000 miles in 1970 and 17279 in 1972. Table 12 summarises the results, and demonstrates the steady improvement attained.

Some rivers have changed spectacularly. The Thames, from Chelsea to Gravesend was deoxygenated and largely lifeless in summer in the late 1940s and early 1950s: by the late 1960s when new sewage treatment works were in operation it never lacked oxygen and at latest count now supports some 80 species of fish, including migratory forms like sea trout which are beginning to enter the river and penetrate the tidal zone (Figure 27: Royal Commission, 1974; Wheeler, 1969). On the other hand, the statistics do to a degree disguise the position. The worst-polluted river reaches tend to be those lowest down, and these are the reaches where the volume of flow is greatest and the largest bankside towns and cities are sited. This is evident in Figure 28 which maps the location of Class III and IV in 1972. Were the statistics arranged in terms of volume, or of human exposure, they would therefore look less encouraging. Many of the estuaries, like those of the Tyne, Tees, Mersey or Trent are in a heavily polluted condition, and moreover sustain inputs from industrial wastes as well as sewage: they thus have problems of toxicity as well as deoxygenation. Metals discharged in such situations are often accumulated by organisms: in the Bristol Channel limpets and other shore molluscs have been found to contain 80 ppm cadmium (wet weight) while oysters in Cornwall

Table 12. *River pollution in England and Wales*

Category	1958 Miles	%[b]	1970[a] Miles	%	1972[a] Miles	%
Non-tidal rivers						
Class I	14603	72.9	17000	76.2	17279	77.4
Class II	2865	14.3	3290	14.7	3267	14.7
Class III	1279	6.4	1071	4.8	939	4.2
Class IV	1278	6.4	952	4.3	832	3.7
Tidal rivers						
Class I	720	40.7	862	48.1	880	49.4
Class II	580	32.8	419	23.4	414	23.2
Class III	250	14.1	301	16.8	253	14.2
Class IV	220	12.4	209	11.7	236	13.2

From Royal Commission (1974).
[a] The total mileage recorded is higher in 1970 and 1972 because the surveys in these years were more accurate than in 1958.
[b] Figures are a percentage of the total mileage of the rivers in the survey.

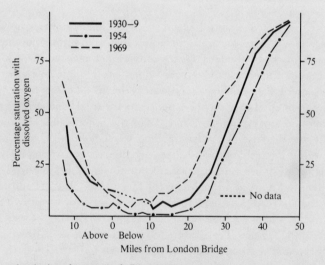

Figure 27. Analysis of water of River Thames. Percentage saturation with dissolved oxygen at high water. Average July–September. From Royal Commission (1971).

Figure 28. Sketch map showing, as heavy lines, the continuous stretches of British rivers classified as poor (class III) or grossly polluted (class IV) in the 1972 survey. From Royal Commission (1974).

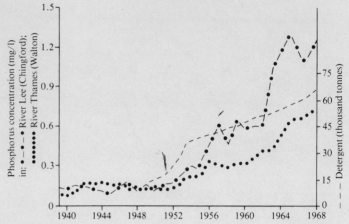

Figure 29. Comparison of phosphate content of two rivers with annual consumption of detergents. From Royal Commission (1971).

are locally discoloured by copper from mining. The Royal Commission on Environmental Pollution, in a recent study of estuaries (1972b; and Porter, 1973), considered that they provided some of the most unsatisfactory examples of pollution in Britain today.

Moreover, while the general trend is towards inprovement, and the classic effect of sewage waste are on the decline due to new sewerage schemes (£1300 million was spent on sewage treatment in the 5 years from 1970), along with parallel declines in some well-established nuisances like toxic industrial wastes, not all trends are equally encouraging. Detergents provide a case in point. In the post-war period, new synthetic detergents came into increasing use, and brought real benefits – but the products did not break down readily in the environment. They therefore interfered with the working of sewage treatment plants, and also created foam on many rivers: in the north with its textile industry, there were notorious examples, the most celebrated at Castleford in Yorkshire, where a weir in the town centre created foamdrifts which were sometimes blown by the wind into the streets to hazard traffic as well as annoy people. Following discussions and trials, a voluntary agreement with industry led to 'soft' or biodegradable detergents coming into domestic use, and the sewage treatment and foam problems largely subsided (Standing Technical Committee on Synthetic Detergents, 1957–77). But the breakdown of these detergents produced phosphate – a nutrient capable of promoting excess plant growth. Some rivers, like the Lee, can be largely composed of treated sewage effluent (up to 50% of the dry weather flow of the Lee at Broxbourne bridge in summer can come from the Rye Meads works), and it is not surprising that phosphate levels have risen (Figure 29), and that

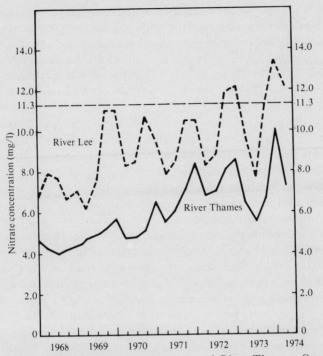

Figure 30. Nitrate concentrations in River Lee and River Thames. Quarterly averages. From Royal Commission (1974).

the Metropolitan Water Board has had difficulties with algal blooms in their reservoirs lower down the valley. In Britain generally these problems of 'eutrophication' have not been so acute as in the United States, because there are few large lakes low down our river systems, but the case for estuarine impoundments in Morecambe Bay or the Wash must depend on the quality of the water retained, and clearly this, taken from the lowest reaches of the rivers, could give rise to real problems with algal blooms and deoxygenation.

Levels of nitrate – also important in promoting plant growth – have also been rising. Probably domestic sewage and run-off from farmland are both sources. Nitrate poses another problem besides its growth-promoting tendency. In bottle-fed infants it can cause a condition of methaemoglobinaemia ('blue baby'): the World Health Organisation has laid down a standard of 22.6 mg N/l (as nitrogen) for public water supplies, and half this concentration is regarded as the desirable upper limit. In 1974 the levels in the Lee and Thames were above this half-way figure of 11.3 mg N/l for short periods and abstraction was stopped. As Figure 30 shows, the trend continues and unless it can be brought under control, special water supplies may be needed for infant feeding: a costly and uncertain business which has already been resorted to on a small scale in north Norfolk and Lincolnshire where high nitrate levels have been found in waters taken from wells in gravel beds. Worry has also been caused because nitrite in gut is alleged, under certain conditions, to combine with other substances to form potentially carcinogenic nitrosamines, though the extent to which this is a real problem is uncertain.

Waterborne lead poses a rather different problem. Natural waters normally contain very little of this metal, except for streams draining orefields and mines, but it has been used extensively in water pipes (as the term 'plumbing' implies) because it is soft, easily worked, but durable. Unfortunately lead can be dissolved by some kinds of water (usually acid and termed 'plumbosolvent') and a recent survey (DOE, 1977c) showed that a significant number of households in Britain can have water issuing from the taps, especially first thing in the morning, with lead concentrations in excess of the World Health Organisation recommended limit of 0.1 mg/l. Such concentrations could result in a lead intake from drinking water comparable with that the average person normally gets from food, and raise total intake to an undesirable level. While the trend in this form of exposure to lead is undoubtedly downwards, as older houses with lead pipes and cisterns are demolished, and modern plastic and copper waste systems come into use, and while the evidence of hazard to man is ambiguous (Shaper, 1977), the survey has naturally caused concern: the phenomenon illustrates well how water that is naturally safe can become less acceptable because of its manner of distribution or use.

There are other fresh-water pollution problems. Oil, seeping from soil or entering drainage systems through spillages, has also caused a mounting number of incidents. It interferes with sewage works and is a disamenity and nuisance. Herbicides at low concentrations, mercury from bulb-dipping and seed potato treatment, and some other industrial releases have shown local upward trends. Even if effluent is treated to high standards it is impossible to remove all traces of drugs (excreted by people as well as discharged in effluent from the pharmaceutical industry) and potentially hazardous chemicals, and the long-term effects of traces of pollution like this under circumstances where water is increasingly being re-used remains uncertain. It is clear that demand for water in Britain and similar countries will probably go on mounting and most will have to come from the rivers. Already these have less capacity to receive and dilute sewage effluent than the original Royal Commission standards proposed. As abstractions and discharges mount (for what goes out generally comes back) the proportion of effluent to 'unused' water must rise. The only way to respond is through increasingly high standards for discharges.

The water situation illustrates the danger of taking things for granted.

In the past, the capacity of rivers and estuaries to dilute and disperse wastes was assumed. But rivers are narrow and have limited dilution capacity, and estuaries are not the open, rapidly changed water bodies one imagines: their waters often move up and down with the tide, gradients of salinity forming invisible barriers. In the past, the waterborne waste-disposal system grew first from the need for cleaner streets, and for domestic hygiene: it established a throughput of nutrients from farms to consumers in the cities and thence to rivers and sea which is wasteful and causes environmental problems. Although a sewerage authority has the power to control discharges to its system, most authorities have under-standably preferred to receive metal-bearing and other industrial wastes into the sewers for central treatment rather than have industry discharge them directly and untreated into the environment. But the result is a dilution and mixing of intractable materials which are then very difficult to treat. Nowadays, more and more of these wastes are being treated at the point of origin where they are more concentrated. Only a partly purified effluent is allowed beyond the 'factory fence'. This practice is more efficient and simpler, and may also be cheaper overall since each industry is likely to be equipped to deal with the kind of wastes its processes create.

4.6 Changes in the sea

4.6.1 Oil at the air–water interface

Oil is the most widely publicised of marine pollutants, highlighted by dramatic incidents like the wrecks of the *Torrey Canyon* in 1967 (Cabinet Office, 1967), the *Metula* in 1974 (Baker *et al.*, 1976) and the *Arrow* in 1970 (Operation Oil, 1970) and blow-outs like that at the Ekofisk Field in the North Sea in April 1977.

The most obvious effects of oil at sea are physical. It spreads at the air–water interface. Its best-known hazard is to seabirds and wildfowl because it clogs their feathers, reducing the amount of air trapped in the plumage and destroying buoyancy and waterproofing. It also blankets shores and has effects on intertidal life discussed in Chapter 5: as a result there is substantial loss of amenity and damage to tourism.

The scale of oil pollution of the world ocean has been disputed. Figure 31 suggests that it is especially a problem of shipping lanes. In some areas the black band of the lichen *Verrucaria*, which grows in the splash zone of rocky shores, is said to have been mistaken for a tarry oil scum, leading to exaggeration of estimates of pollution. However, the amount of oil entering the sea annually has undoubtedly risen substantially in past decades and the trend continues (Wardley Smith, 1976). The pathways are more complicated than is generally appreciated. Preoccupation with oil tankers tends to divert attention from the equally large contribution from

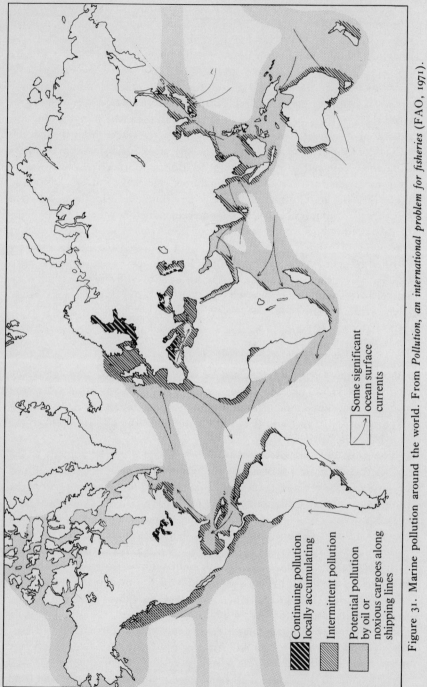

Figure 31. Marine pollution around the world. From *Pollution, an international problem for fisheries* (FAO, 1971).

Continuing pollution locally accumulating

Intermittent pollution

Potential pollution by oil or noxious cargoes along shipping lines

Some significant ocean surface currents

Table 13. *Estimated direct petroleum hydrocarbon losses (millions of tonnes) to the marine environment (not including airborne hydrocarbons deposited on the sea surface)*

	1969	1975 (estimate) (Minimum)	1975 (estimate) (Maximum)	1980 (estimate) (Minimum)	1980 (estimate) (Maximum)
Tankers	0.530	0.056	0.805	0.075	1.062
Other ships	0.500	0.705	0.705	0.940	0.940
Offshore production	0.100	0.160	0.320	0.230	0.460
Refinery operations	0.300	0.200	0.450	0.440	0.650
Oil wastes	0.550	0.825	0.825	1.200	1.200
Accidental spills	0.200	0.300	0.300	0.440	0.440
Total	2.180	2.246	3.405	3.325	4.752
Total crude oil production	1820	2700		4000	

Source: Study of critical environmental problems: Man's Impact on the Global Environment, ed. C. L. Wilson and W. H. Matthews. Cambridge, Mass.: MIT Press.

other ships and from oil wastes entering the sea from rivers and from shore-based industry (Table 13). Substantial amounts of oil are believed also to reach the sea as tiny airborne droplets.

This upward trend has been understandably unwelcome. In recent years there have been substantial national and international steps towards control, described in Chapters 8 and 9. The effect has been to contain, or even reduce, the amount of oil entering the sea from tankers and other vessels. Action under agreements like the Paris Convention (Chapter 9) should reduce waterborne emissions of oil from land-based sources in north-west Europe, where national regulations are also becoming tighter. On the other hand the increasing exploration and development of undersea oilfields, including those in ice-beset waters off Labrador, Newfoundland and Alaska bring new risk of accidents. DOE (1976d) estimated that there was a 50:50 chance of a blow-out in the North Sea before 1982: this prediction was validated by the Ekofisk Bravo incident in April 1977. New precautions will undoubtedly reduce the probability of future accidents, but cannot wholly eliminate them. It is thus difficult to predict the likely trend in oil pollution of the oceans, despite all the efforts being devoted to its containment.

4.6.2 Changes in the chemical composition of the seas

As Chapter 3 pointed out, the oceans are the chief global 'sink' where pollutants accumulate. The chief causes of concern apart from oil have been

Table 14. *Man-induced rates of mobilisation of materials which exceed geological rates as estimated in annual river discharges to the oceans*

Element	Geological rates[a] (in rivers) (Thousand tonnes per year)	Man-induced rates[b] (mining)[c]
Iron	25000	319000
Nitrogen	8500	9800
Manganese	440	1600
Copper	375	4460
Zinc	370	3930
Nickel	300	358
Lead	180	2330
Phosphorus	180	6500[d]
Molybdenum	13	57
Silver	5	7
Mercury	5[e]	5
Tin	1.5	166
Antimony	1.3	40

Source: Study of Critical Environment Problems: Man's Impact on the Global Environment, ed. C. L. Wilson and W. H. Matthews. Cambridge, Mass.: MIT Press.

[a] Bowen 1966 – [b] United Nations, Statistical Yearbook – [c] 1967 data for mining except where noted – [d] Consumption. [e] Not including outgassing from coastal rocks, which may release as much as 150000 tonnes of mercury a year.

persistent and highly toxic substances liable to accumulate in marine life, and Chapter 2 lists many of these. Trends in the emission of heavy metals have undoubtedly been upwards in recent decades, although the release of many of them to the ocean by man is likely to be less than comes from the natural weathering of the rocks. Perhaps 5000 tonnes of mercury, for example, are emitted to the sea by man each year (world usage is currently about 9000 tonnes/year) whereas natural erosion contributes some 5000 tonnes/year more and the de-gassing of the earth's crust may release 150000 tonnes, much of which ends up in the sea (for further recent figures see DOE, 1977a). Estimates for other substances which man may release on a scale comparable with that in nature are given in Table 14. Alongside these natural substances, we have been discharging to the sea a widening range of synthetic chemicals, mainly as industrial wastes. Some have been dumped directly in the oceans (probably only about 10% of the total input, but with a disproportionate share of the most toxic substances). Others enter from the land via coastal outfalls, drains and rivers – and in a few places through the tipping of waste onto the foreshore.

Materials such as metals, organochlorine pesticides and PCBs that are

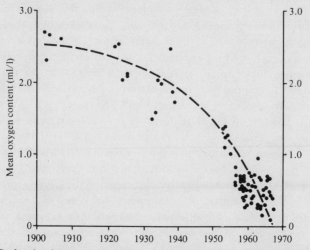

Figure 32. Reduction in oxygen concentration in deep water, Landsort Deep, Baltic Sea. From Royal Commission (1971).

persistent and accumulate in living organisms attain high local concentrations – especially in confined waters flanking industrial zones, like the Irish Sea, North Sea, Baltic or North American estuaries. There is insufficient evidence to make statements about trends (although there is some indication that mercury levels in fish may always have been near today's values). What is important is that high concentrations and possible hazards remain localised: they are 'hot spot' problems of estuaries and inshore seas rather than threats to the stability of the world ocean (Figure 31). And action has already been taken to curb dumping and discharges from the land in north-west Europe (Chapter 9).

4.6.3 *Changes in British inshore seas*

Even in the seas around Britain, scene of many classic studies of marine biology, there are few reliable baseline measurements from which to judge trends (Royal Commission, 1972). We can be sure, however, that no areas of sea around Britain approach the deoxygenated condition of the Baltic (Figure 32). Some, however, have local oxygen depletion and enhanced levels of some metals in benthic sediments, molluscs and inshore fish. A recent survey showed that fish caught in the eastern Irish Sea (east of a line from Anglesey to St Bee's Head), in the Thames estuary, and on part of the Kent coast had mean levels of mercury above 0.5 ppm (MAFF, 1971, 1973). Probably this mercury came from industrial discharges and one chlor-alkali plant releasing effluent into Morecambe Bay immediately took voluntary control measures which reduced its releases to below one-tenth of their previous level: a pesticide plant in the same area, which had also been using mercury, closed shortly before the survey discovery. The Irish Sea is also relatively high in other metals and in PCBs, and it was probably these that contributed to, or triggered, the deaths of 12000 seabirds in 1969 (as described in Chapter 5). A recent working party on sewage sludge dumping in Liverpool Bay, in the eastern Irish Sea, revealed in its report, aptly termed 'Out of sight, out of mind' (DOE, 1972b), that despite the large amounts of dumped material (100000–250000 tonnes per annum) effects on oxygenation were local and never led to depletion while effects on the flora and fauna of the sea-bed were localised and less than those caused by pollution borne by direct drainage from the land. A rather similar pattern holds where sludge is dumped in the Clyde (where however the sediments can contain over 300 ppm of lead) and off the Thames. But the 'soup' of discharges of all kinds in the Irish Sea is presumably the source of the residues of PCBs, metals and other contaminants carried in its seabirds, and this sea is likely to be one of the most contaminated in western Europe, apart from the shallow and polluted waters trapped among the islands and shoals of the Netherlands and supplied with an even nastier 'soup' by the inflowing Rhine (Wolff, 1978).

A recent study of the North Sea by ICES (the International Council for the Exploration of the Sea) was, however, broadly reassuring (ICES, 1969). It demonstrated that fish populations showed no signs of damage from pollution despite the fact that the nursery grounds for many species are inshore, in estuaries and bays, where pollution was highest. A recent further analysis (Sibthorp, 1975) did not dissent, but emphasised the many gaps in our knowledge even of this much-studied sea. ICES is monitoring the condition of European shelf waters in collaboration with the Commission of the Oslo Convention, and will therefore be able to watch for general trends, including any resulting from pollution with hydrocarbons from the North Sea oilfields. In Britain, monitoring of mercury in fish, metals in shellfish, and PCBs and organochlorine pesticides in seabirds and their eggs reveal no clear trend, but there are signs of a decline in organochlorine burdens in some coastal birds. On the other hand, oil pollution of beaches and inshore seas has almost certainly increased in recent decades and more sewage, sewage effluent and power station cooling water has been released. Some estuaries, in particular, remain in a seriously polluted state (Royal Commission, 1972; Porter, 1973).

There are definite local improvements to set against deteriorations. Around the oil refinery at Fawley about a quarter of a square mile of salt marshes was killed by repeated contamination with low concentrations of oil, over the years up to about 1970. Improved waste treatment has now reduced the oil level in the effluent to around 5 ppm, allowing the vegetation to recover (Dicks, 1976). At Milford Haven, intertidal flora and fauna are affected in the immediate vicinity of outfalls, but the gradient is a sharp one and little effect is detectable at 100 m or more (Baker, 1976). In Orkney and Shetland effluents to be discharged at Flotta and Sullom Voe are likely to have average oil contents around 10 ppm, and these should not create significant 'hot spots' of marine damage. Shore-based oil installations need not, therefore, be a source of major pollution if properly designed and operated, and guarded against accidents.

The curbs now placed on dumping of toxic wastes at sea under the Dumping at Sea Act, 1974 (which allowed the United Kingdom to ratify the Oslo and London Conventions) will help to prevent a further decline in the conditions of inshore waters. Curbs on discharges to the rivers, estuaries and coasts, under the Control of Pollution Act, 1974 and the Paris Convention, and the increasing recovery of toxic materials at source by industry before they enter the sewer (and hence end up in sewage sludge) will do more.

4.7 *Changes on land*

There are no examples of global land pollution: the nearest are those resulting from the long-distance dispersal of airborne materials like lead, already described. As an example of the more localised pattern typical of most industrial countries, the situation in Britain will be used as it was for fresh-water pollution. Most land pollution in Britain comes from three sources: fall-out from the air (e.g. of lead or other metals), dumping of wastes and the side-effects of herbicides and pesticides. Airborne fall-out has in the past been greatest around industrial sites (e.g. in the lower Swansea valley, where smelters brought zinc levels in soil up to 20000 ppm and almost created extractable ore (Weston, Gadgil, Salter and Goodman, 1965). Dumping of wastes has been practiced for centuries – indeed millennia – and ancient rubbish dumps are among the archaeologist's most fruitful study areas. Inevitably, the volume and complexity of such wastes has increased in the modern industrial era, but most dumped materials are innocuous, and so long as they do not directly foul water supplies, they will have little effect on human health and well-being. Modern methods of burial of waste under soil have certainly remedied the worst disamenity and pest infestation problems of old uncontrolled tipping.

In recent years, however, there has been considerable concern over the unauthorised dumping or 'fly tipping' on land of toxic wastes and legislation has been introduced in 1972 and 1974 to deal with this. All wastes containing the materials scheduled under this legislation must now be deposited on sites approved for the purpose by local authorities. The most noxious materials may need treatment before disposal. Now that the sea is closed also to the dumping of many toxic materials, their safe disposal is increasingly difficult and chemical reclamation from concentrated waste, or the combination of the materials in an inert form that will not allow leaching to contaminate watercourses can be expected. The chief problem in future may come when old dumps, including unauthorised and unrecorded tippings of noxious substances, suffer the corrosion and disintegration of containers and the release of their contents.

Pesticides have been deliberately applied to the land environment for centuries. The earlier mixtures containing lime, copper, or arsenic, included natural materials and had locally toxic effects on a range of organisms. After 1945 the new generation of organochlorine pesticides came into increasing use, and their side-effects on predatory birds and mammals soon also became apparent, as Chapter 5 explains. Since then there has been voluntary control under the Pesticides Safety Precautions Scheme, operated under the guidance of the Advisory Committee on Pesticides and other Toxic Chemicals, and the development of less persistent and more specific substitute materials (subjected to increasingly rigorous screening before introduction) has proceeded apace. Usage of persistent organochlorines in England and Wales fell from 460 tons in 1963 to 300 tons in 1967 and 250 tonnes per year in 1970–2 (Royal Commission, 1971, 1974) (in the United States the comparable figures are 60000 and 44000 tonnes). Dieldrin and aldrin, two of those which caused most losses to wildlife, have been phased out. The targets whose deaths caused most concern a decade ago – eagles, peregrine falcons and other bird predators – are now increasing again and the reproductive success of herons, invaluable as indicators, has improved. In a similar way, the PCBs, often compared to the organochlorine pesticides although in fact different in their basic structure and used industrially rather than in agriculture, are not now supplied in Britain for uses that might lead to the contamination of the environment. Their residues in wildlife should soon start to fall. The pesticide story has, therefore, a trend of success to record. However, it is not entirely one way. In 1973 there were mysterious deaths among tomato crops in Essex, ultimately traced to very low levels of a herbicide which was discharged into the Ouse catchment and transferred to the headwaters of the Suffolk and Essex rivers in a scheme designed to augment their flow by pumping water from low down the Great Ouse system. In 1974–5 there were unexpected deaths among wild geese in Lincolnshire, traceable to a new pesticide, trithion or carbophenothion, introduced as substitute for the DDT group. Even with the most stringent screening, it is not always possible to be sure that an agent of this kind will not prove toxic to some unexpected target under some circumstances.

4.8 Radioactivity in the sea and on land

Radioactive materials have been stringently controlled from the beginning of their non-military use, and especially in recent years when the potential hazards of even their earlier employment in medicine have become apparent. Industrial processes using radioactive materials and emissions to the environment are very tightly controlled. It is a well-known fact that the level of discharge of the persistent isotope of ruthenium, Ru 106, from the fuel treatment works at Windscale, is set so as to avoid conceivable hazard to people in South Wales who eat the seaweed *Porphyra* (laver), which accumulates this isotope. The need for such a precaution, despite the apparent dilution capacity of the sea, has been confirmed by research that showed how the waters into which Windscale discharges circulate along the British coast like an inshore river, hugging the shores as they flow northwards about Scotland and then south into the North Sea.

On land, one problem of radioactivity is the handling of the wastes which are created in all nuclear power stations (Royal Commission, 1976b; Command 6820, 1977). The trend to nuclear fuels and away from fossil materials like oil or coal has been welcomed as a way of reducing air pollution. Conversely, other problems of pollution prevention are created in consequence. At present, the least radioactive waste, of shortest half-life, is dumped at sea in deep water under controls and procedures agreed by the United Nations Scientific Committee on the Effects of Atomic Radiation (UNSCEAR) and the International Atomic Energy Agency (IAEA). More persistent and active residues are stored as a sludge in thick concrete tanks. The material is at present only being accumulated slowly and there is ample safe storage space, but some concern has been voiced over what will happen if fast-breeder reactors which produce some 'hot' waste with a 'half-life' of tens of thousands of years are adopted (Royal Commission, 1976b). Present proposals favour storage, perhaps cast into thin vitrified 'biscuits' in salt mines, stable granite rocks or other deep and shielded vaults. The volume of the material presents no great problem: the chief need is safety from earthquake, flood or other upheaval. Research into the most appropriate storage sites and methods is going on (see Cohen, 1977 for a good general review of this issue).

4.9 The scale of change

Only a few pollutants appear to be potential global threats. A number are or have been until lately, on a rising trend. Some have posed serious local problems. In most, if not all cases, where genuine hazard has been proved, some action to curb or reverse the trend has been taken. This analysis suggests therefore that those pollutants still on a rising trend in the environment are those whose hazard has not yet been established, or not been established long enough for control measures to bite. Admittedly, global monitoring is not good enough at present to define trends with high precision or define the variation in their rate and direction over the world: unless it is improved, we may remain uncertain who is winning or losing where and how fast. But the analysis may suggest that the pollutant to be concerned about is the one whose hazard has not yet been recognised. Since most hazards have been demonstrated because of events where the concentration of the pollutants concerned is highest – in local 'hot spots' – it also should follow that action is most advanced in densely peopled and environmentally aware countries, like those of Europe or North America. It may also be predictable that more action will be taken where the localised problems occur frequently – a phenomenon sometimes called 'the measles effect'. An analysis of the British experience, from one of the world's smallest, first industrialised and densely populated states, provides some evidence for this prediction. It seems generally true that the trends of many pollutants in Britain, as at global level, are in the right direction – downwards. But there are exceptions, like nitrate in fresh-waters and perhaps PCB residues in birds, which have not yet had time to respond to recent controls. The possibility that levels of new substances whose hazard has not been defined are still rising – as those of chlorofluorocarbons are – remains and demands a wide monitoring and evaluation activity, linked to an effective machinery for administrative and legal action. Linkages between such mechanisms are needed internationally if the materials in question are widely used, widely traded or can contaminate the air or ocean over whole regions, and what is being done is outlined in Chapter 9.

Chapter 1 emphasised the importance of biogeochemical cycles in the ecology of the world. While most of the recorded effects of pollution on land are due to direct contamination with toxic substances, side-effects on the soil ecosystem can be expected through the use of herbicides and pesticides whose action is not specific to the targets against which they are applied. Similarly, rates of decomposer processes can predictably be altered, with effects on the return to air and water of soluble or volatile products. Such effects today are likely to be on a small scale and subtle. There is certainly no evidence that major cycles of oxygen, carbon, nitrogen, phosphorus, sulphur or other key nutrients have been so altered by pollution as to threaten life on a wide scale. All investigations of possible declines in atmospheric oxygen levels, for example, have tended to the conclusion that these have not changed since the first reliable measurements (e.g. SCEP, 1970). Human activity has none the less brought about major alterations in some of those cycles (Svensson and Soderlund, 1976; SCOPE, 1977). For example, the dominant effect of human activity on the nitrogen cycle is to increase the fixation of the gas from the atmosphere. This occurs by combustion, which creates nitrogen oxides (passed back to the atmosphere and possibly able to affect ozone levels in the stratosphere) and nitric acid (deposited on the ground and causing acidity). Fertiliser manufacture and new agricultural methods, employing plants that fix atmospheric nitrogen, also augment nitrogen levels in soil and drainage, and some of this finds its way back to the atmosphere as nitrogen oxides. Levels of use of nitrogen fertilisers continue to rise, suggesting that these changes are likely to be accentuated.

The separation in space of the food producers and food consumers has had a particular impact on the phosphorus cycle. There is a massive flow of this element from land to city, and due to the features of our sewerage system, thence to the sea. About 14 million tonnes of phosphorus are lost from the land each year. Phosphorus is relatively stable in the soils being scarcely leached out in drainage water but it is only slowly replenished from the air through rainfall. Probably only 0.1 million tonnes a year return in this way from sea to land: most natural replenishment comes from rock weathering (Institute of Ecology, 1971; SCOPE, 1977). Man has probably doubled the global flow-rate of this element, making good the massive drain to the ocean by mining reserves of phosphatic rocks (which some calculate will run out in a century or so). This influx seems to have no detectable impact on the ocean: the significant implication for man is that we remain dependent on replenishment of soil phosphorus to sustain agriculture, and if reserves of easily-won rocks are depleted we may need to modify the part of the cycle under our control, for example by reducing losses in sewage.

The sulphur cycle has largely been altered by the release of sulphur dioxide when fossil fuels are burned. In the natural system about 50 million tonnes/year entered the air (Table 11), this emission being exactly balanced by deposition. Man's effect has been to add about 65 million tonnes to the annual emissions, and cause a compensating increase in deposition, largely over land (49 million tonnes) but partly to the sea (16 million tonnes).

It is important to stress two things. First, that these modifications to the biogeochemical cycles are not uniform over the world: our environment is highly variable and regional differences are marked and especially related to the pattern of distribution of industry and intensive agriculture. Second, few geochemical cycles have escaped some alteration by human action, partly because ecological changes brought about by man naturally affect the flux of all those substances that pass through or form part of living organisms and partly because the cycles interact and where climate may be affected there is a general impact on the processes throughout the biosphere. This increases the need for scientific understanding of these systems and their variation and the causes of their determinant processes.

5. The effects of pollution on targets

5.1 *Effects and targets*

The interaction of target and pollutant results in some kind of effect, and it is the nature, scale and significance to man (including the cost) of this effect that is at the heart of the whole pollution value judgement. The prediction of how effects will change with time is an equally essential part of the process of risk estimation.

We traditionally establish a rank-order of targets. We are most concerned to protect man: least concerned with organisms like bacteria which many people find aesthetically unappealing, and with others like rats which are pests and vectors of disease. The traditional rank-order, which is in itself an intuitive risk-assessment, can be represented, in order of decreasing concern:

Man → Domestic → Crops and → Most wildlife → Pests and
     livestock   structures   and amenity   disease vectors

Different people and communities may well reverse parts of the sequence, for example ranking wildlife and amenity as of higher importance than some structures (especially ugly ones).

In assessments like this we commonly discriminate between *acute* damage, defined as an adverse effect on a target that follows fairly immediately from exposure to a pollutant and is generally clear-cut, often fatal, and rarely reversible, and *chronic* damage that commonly develops long after exposure, or after prolonged exposure to low doses, is less clear-cut, and may well be reversible (Saunders, 1976). Generally, acute damage is regarded as more serious than chronic damage, and by implication at least, a rank-order of effects paralleling that of targets is often recognised, although far from universally, again in order of decreasing concern:

Acute toxicity → Chronic effects → Impairment of → Impairment
or death   ultimately caus-   growth or   of behaviour
                  ing death   function

The two rankings together may be used as a matrix, as an aid in judging the seriousness of an effect, and this is illustrated in Table 15. Just how

Table 15. *Matrix setting out the rank-orders of concern for living targets and scales of damage*

Effects	Target				
	Man	Domestic livestock	Crops and fisheries	Wild life	Pests and disease vectors
Acute toxicity and death	+++	+++	+++	++	AAA
Chronic damage and ultimate death	+++	+++	++	++	AAA
Impaired growth, function or reproduction	+++	+++	++	+	AA
Impaired behaviour	+++	+++	++	+	A

Key: +++, effects wholly unacceptable: corrective action required as soon as detected; ++, effects unacceptable on any significant scale: corrective action required as soon as scale of damage established; +, effects unwelcome, requiring investigation and monitoring with corrective action if on a significant scale and if resources allow; AAA, effects beneficial, but monitoring still required to record scale and nature; AA, A, lesser effects, probably still beneficial, but scale requires assessment and monitoring desirable.

much damage to particular targets is tolerated depends on economic and social circumstances, but generally communities are intolerant of any impairment of human performance, even in a minority of the population, and impaired behaviour in livestock or pets is also generally unacceptable. Economic loss occurs if there is more than minor impairment of crop growth and function, and communities act accordingly to keep such effects within bounds. We seem more willing to accept some mortality among certain kinds of wildlife, and we may positively welcome pollution that prevents mildew in our roses or kills midge larvae in the pond in the park.

This simplistic analysis can, however, be misleading. First, it overlooks the fact that a pollutant commonly affects more than one target in the environment at a time: man is not the only organism affected by smoke or sulphur dioxide in the air of an industrial city. Second, it overlooks the fact that man may be as much affected circuitously, by effects which undermine the ecological systems on which the fertility of the soil or productivity of the seas depend as by a direct hazard. It is useful to

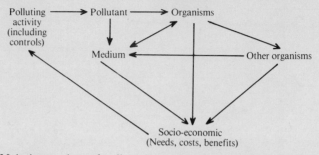

Figure 33. Main interactions of pollutant, target and socio-economic system.

discriminate in this context between a *primary target* or *impact*, where we are concerned with the direct consequences of a pollutant, and a secondary target or impact involving circuitous interaction via other organisms.

The basic model for the effects of pollutants can be represented by Figure 33. A pollutant may modify the physical and chemical properties of air, water or soil and affect organisms through habitat changes (Chapter 4). Or it may pass through the media with little direct effect, to influence organisms and through their changes, other organisms. There is feedback to the media through chemical changes linked to the biological variables, whether mediated via the decomposers and the recycling of elements or through such factors as the deoxygenation of water. Virtually all the processes in Figure 34 are liable to be perturbed by pollution.

It is possible to classify target: pollutant interactions according to their position in such a sequence, and according to the basic nature of the pollutant, and thus define primary, secondary and tertiary impacts (Figure 35): several are commonly present. Primary impacts of pollution on organisms involve changes in basic biochemistry or energetics, and hence production and survival (measured as growth, longevity and reproduction) whereas secondary impacts operate through altering food supply or the balance of competing species. Thus primary effects tend to be physiological, and secondary impacts ecological.

It is implicit in this system that there is something of a 'cascading' effect. The basic impact of any pollutant is to affect the rate and direction of chemical or physical processes. Physical pollutants acting directly on organisms may alter the rate of temperature-dependent reactions, the interactions of enzymes and substrate, or the ion balance in body fluids and across nerve membranes. Chemical pollutants impinge directly on biochemical and biophysical reactions like these. If those biochemical and biophysical effects are relatively minor, they remain within the resilience of the organism: that is, like most minor natural fluctuations in the composition of food or the salinity of media in which animals live or plants root, their impact can be cushioned by its homeostatic mechanism. There

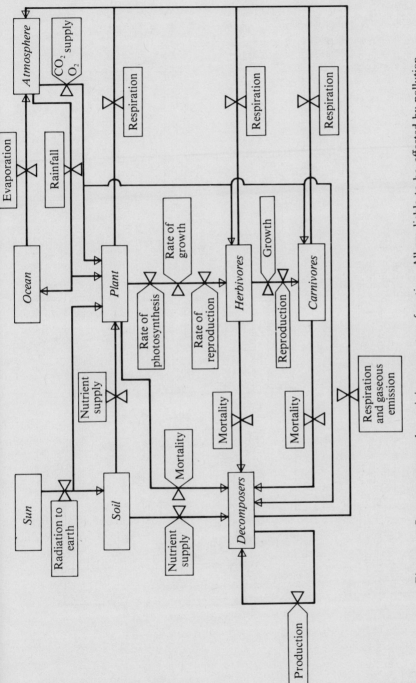

Figure 34. Some processes determining ecosystem function. All are liable to be affected by pollution.

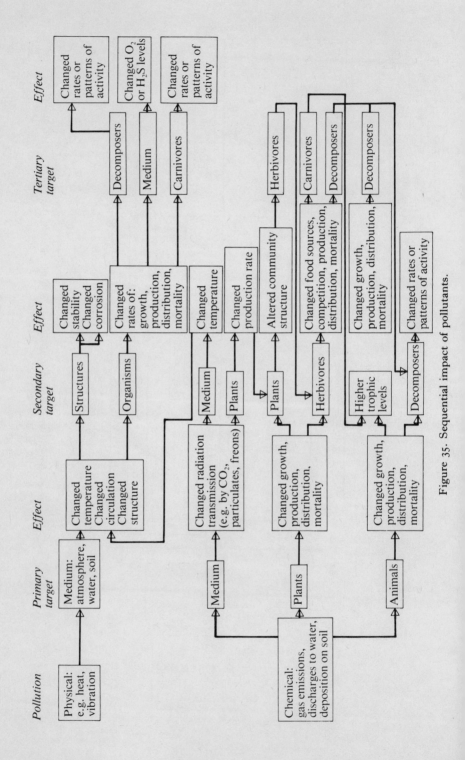

Figure 35. Sequential impact of pollutants.

may thus be detectable biochemical changes and enhanced energy consumption, but no evident physiologically significant response.

As exposures mount, the biochemical disturbances may spill over to perturb the working of organ systems. Even when physiological symptoms are detectable: when growth rates decline or pigmentation or reproduction are altered, there will not necessarily be evident effects on overall vigour. But at higher or more long-enduring exposure levels there are likely to be sufficient effects for at least the most susceptible members of a population to show impaired competitive vigour, diminished production of young or shortened life, and as this kind of impact mounts, effects commence at the population level: as the susceptibility or resilience of species is likely to vary this may in turn lead to alteration in the composition and balance of communities, thus affecting the ecosystem level as the last 'overspill'.

A model of such a system, in crude terms, is given in Figure 36. It provides both for homeostatic capacity at each level and for the fact that a critical parameter is the rate of 'input' of the pollutant and the rate at which it can be rendered harmless by direct mechanisms of detoxification or excretion (at the biochemical level), replacement or new growth of damaged tissue (at the organ level), recruitment of otherwise surplus individuals into the breeding stock (at the population level) or a changed species balance that yet sustains the overall physiognomic characteristics and structure of the system as a whole (at the ecosystem level).

A crude model of this kind is not worth elaborate refinement, but it can also describe another feature of the system: variation in resilience between individuals and at different levels of the system. Allowance must also be made for the evolutionary response of species through recombination, mutation and natural selection. For example, certain strains of plant found on soils with high mineral levels are mineral tolerant, and this capacity is fairly rapidly developed in a population (Wainwright and Woolhouse, 1975). Metal-tolerant populations of invertebrate organisms are likewise well known. In many cases such genetic changes probably arise through the selection of characteristics already present in the species, but the development of new characters by mutation may also play a part. The 'recruitment' of new genotypes can be catered for as a dynamic development of the model, varying the dimensions of the homeostatic 'reservoirs' and the balancing capacity of the 'detoxification' outlets with time.

This question of level of response, and the ease with which we can detect it is vital in everyday terms. Our basic judgement about whether a pollutant effect is significant commonly hangs on whether there are evident effects on whole organisms. If people are dying, concern is natural. But so it is if people are sick, crops producing less or wildlife failing to breed. Such effects are the cumulative consequence of bio-

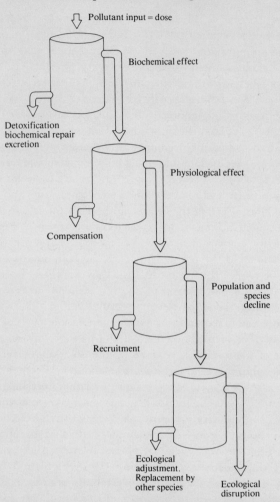

Figure 36. Crude model of 'cascading' effect of pollution, as a series of reservoirs. Effects do not spill over from one level to another unless inputs exceed capacity of relieving systems depicted on left of each reservoir. Variation between individuals can be thought of in terms of variation in capacity of these systems, which may also be increased by evolution, producing more resistant organisms.

chemical and physiological disturbance, progressively departing from normality long before organisms start to die. Such changes, short of the lethal level, are commonly called 'sub-lethal effects'. Many people find this term confusing. It simply means 'effect detected at the biochemical or physiological level, but not manifest in the death of individuals or through changes in populations'. Any disturbance that is big enough to modify the behaviour of the whole body is likely to reduce the efficiency of the

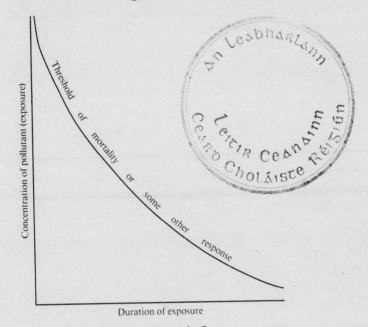

Figure 37. Relationship between exposure and effect.

organism so much that it is less well able to compete for food, less well able to endure a period of adverse climate, or less well able to reproduce. It is likely to be manifest in impaired growth if the stress is such that most of the available energetic intake from food is diverted to the maintenance of basic organic functions. Within a population there will naturally be both genetic variations and variations due to past events – for example in food intake and hence in body energy reserves – and these will be reflected in the response, explaining why the impairment of efficiency can be lethal to a fraction of the organisms under stress while most do not die in any short period of observation as an evident consequence of the pollutant.

It is this marginal impairment of performance that is often called a sub-lethal effect. It does not mean that over long periods of exposure death rates will not be accelerated (Saunders, 1976). The contrast is with acute effects leading to death within short periods, such as those used in laboratory toxicity tests (Chapter 2). The curve relating mortality (or some other defined level of damage) to exposure to a pollutant not unexpectedly follows a logarithmic form (Figure 37). Developed to allow for variability within a population, it becomes transformed to the pattern in Figure 38. When examining the impact of pollutants on organisms, or devising methods to protect individuals or populations, this variability in response according to exposure and individual characteristics must be allowed for.

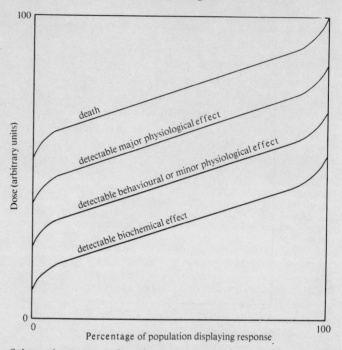

Figure 38. Schematic pattern of variation in response to pollution within a population. The lines represent threshold doses at which the various responses are first detectable.

5.2 Effects of sound, radiation and temperature on living targets

In Section 2.2 it was pointed out that ionising radiation (cosmic, nuclear and X-rays) was to be regarded as a pollutant in its own right, while in Section 4.3.4 and Table 10 the possible effects of aircraft, chlorofluorocarbons and oxides of nitrogen derived from fertiliser manufacture and use on stratospheric ozone concentration were described. The possible depletion of this ozone is a cause for concern because it screens out ultraviolet radiation in the 290–320 nm range (so-called UV-B). All these types of radiation have a primary impact (often called a primary lesion) on DNA (desoxyribonucleic acid) which controls the structure of organisms. If their DNA is damaged, dividing cells may produce abnormal progeny which fail to form the normal harmonious contact with their neighbours, multiply unduly rapidly and produce cancers. Damage to the reproductive organs may likewise cause genetic abnormalities passed on to later generations.

UV-B only affects DNA in the skin of man, and perhaps other species not adequately protected by hair, feathers or pigment, because it cannot penetrate deep into the tissues. But because it affects the basic biochemical control over cell reproduction it is liable to cause both malignant melanoma skin cancers and less serious non-malignant growths. Skin pigmentation is an important defence against this effect: in the United States the incidence of malignant melanoma varies from 4.6 and 4.4 cases per 100000 white men and women to only 0.9 and 0.7 per 100000 in black males and females (age-adjusted data, from CISC, 1976). Table 10 sets out one series of estimates of the likely changes in cancer incidence that might follow various changes in the numbers of high-flying aircraft and chlorofluorocarbon and nitrogen oxide release. To keep the subject in perspective, it should be recalled that exposure levels have been increased up to tenfold in some parts of the United States because of the modern enthusiasm for sunbathing and scanty clothing: simple preventive measures including lotions, opaque clothes and sensible behaviour can greatly reduce risk among the susceptible population (CISC, 1976; DOE, 1976; Machta, 1976). More deeply penetrating ionising radiation is more serious, because it can cause cancers in more vital and sensitive organs and can also modify the genes, causing mutation, abnormal embryos and abortion (see Section 5.2.4). Any such radiation in the general environment, moreover, cannot be detected by the senses or avoided by changes in individual behaviour.

Noise operates in two ways (McKnight, Marstrand and Sinclair, 1974). First, it may damage the hearing, and this has been a long-standing problem in some kinds of industry, where noisy equipment has to be used. The problem is now less acute than it was, due to better machinery and the wider use of protective mufflers (although a surprising number of people are reported still to ignore the protection and incur injury as a result). Likewise, vibration from some kinds of equipment (including some chain saws) is known to be capable of causing injury. Outside working areas, there has been worry over the threat to hearing posed by excessively loud 'music' played in entertainment halls. But in the general environment noise is more of a nuisance and stress than a direct threat to the human physiology. As with other chronic exposures, there is far less information about the extent and severity of these effects, but control measures are being increasingly applied because of the disamenity which noisy aircraft, vehicles, equipment and even entertainment undoubtedly impose on the community at large.

Chapter 4 described how waste heat was inevitably discharged into the environment as a result of energy generation, forming 'heat islands' in the air over cities, and warmer reaches of rivers and coastal seas receiving the outflow from power station cooling systems. At Hunterston in Scotland in 1971 the throughput of cooling water was 91000 m^3/hour, and the returning water was 10 °C hotter than the sea, although it was swiftly mixed (Barnett, 1971). There were consequent changes in the organisms living in the nearby seas: the amphipod crustacean *Urothoe brevicornis*, for example, bred earlier and had a longer growth period than in nearby unheated waters. Raymont (1964) similarly recorded a longer breeding period for the barnacle *Elminius modestus* in heated areas of Southampton Water. Where the water passes into an enclosed basin the effects can be more profound: for example Naylor (1965a, b) recorded at Swansea a discharge into an enclosed dock that was warmed to 5 °C above the previous level so that a sub-tropical barnacle replaced the native British species and various other organisms typical of warm water appeared. Various other warm-water species have appeared in temperate zones under such conditions, including an American clam and a warm-water copepod near power station discharges in Southampton Water. At present, these faunal anomalies are localised, but the possibility of wider biogeographical changes including the appearance of species that become pests cannot be ruled out. Moreover, since the biological processes of all cold-blooded organisms are temperature dependent and raising temperature 10 °C can double the rate of many of them, warming the environment can increase the rate at which nutrients are consumed and if nutrients are not limited, accelerated growth occurs and masses of organic matter pass to the decomposers. The consequent activity of the latter can deplete the medium of oxygen, ultimately creating an anaerobic 'sulphuretum' emitting hydrogen sulphide. Estuaries and confined seas, where warm waters are generally released, are also often polluted with sewage (O'Sullivan, 1971, Royal Commission, 1972b) and increased algal growth and deoxygenation sometimes occur in such areas. On the positive side, it has been suggested that warm water from power stations would be ideal for certain kinds of fish farming, because fish and their food organisms would grow more quickly than in the cold seas outside (cf. Beauchamp, Ross and Whitehouse, 1971).

5.3 Effects of chemical pollutants on animals

5.3.1 Biochemical and physiological effects

There are relatively few well-worked examples of the biochemical effects of pollutants. The best documented are for organochlorine compounds like DDT, organophosphorus pesticides, and certain metals.

Haemoglobin, the pigment that makes blood red and carries oxygen, consists of an active component, haem, bound to a protein, globin. The enzyme delta-aminolaevulinic acid dehydratase (ALA-D) controls one stage in the series of reactions by which haem is synthesised. Its activity is blocked by lead, which also interferes with several other enzymes in the

sequence (haem synthetase, ALA synthetase, uroporphyrinogen decarb-
oxylase and coproporphyrinogen oxidase: Clayton, 1975). In man, ALA-D
activity is reduced when the level of lead in blood reaches around 20 μg/100
ml, which is within what are commonly regarded as normal limits.
Extensive reviews of blood lead levels in peoples remote from industrial
contamination, including Australian aborigines, Kalahari bushmen, New
Guinea tribesmen and remote Amerindian communities have recorded
values from 15 to 25 μg/100 ml (DOE, 1974b). In Europe and North
America, rural communities also have blood lead levels in this range, and
although city dwellers have somewhat higher concentrations, only a small
proportion exceed 30 μg/100 ml. While more recent studies in South
America cast doubt on the earlier figures and suggest that the true blood
lead level in a non-industrial population should be under 10 μg/100 ml, it
remains a fact that a large part of the world's population shows no obvious
sign of lead poisoning yet has blood lead concentrations overlapping into
the range within which haem synthesis can be affected. At 30 μg/100 ml
ALA-D activity is significantly impaired, while at levels of 40–50 μg/100
ml, by no means uncommon in occupationally exposed groups or
inhabitants of contaminated urban regions, the enzyme activity is reduced
to 10–50% of normal level (Hernberg and Nikkanen, 1970).

This might appear a serious hazard since maintenance of adequate
haemoglobin levels is a matter of some importance. But one aspect of the
physiological homeostasis of living organisms is the 'over-design' of many
systems, with excessive capacity and with alternative ways of achieving
something should the normal machinery be blocked. Additionally, or
alternatively, biochemical systems include a repair machinery so that
depleted or damaged enzymes can often be replaced rapidly. Although
ALA-D activity is a very sensitive index of exposure to lead, the clinical
significance of its depletion is therefore far less certain. There is evidence
from research on laboratory animals, moreover, that even after its levels
have been reduced greatly, blood loss can still be made good fairly rapidly.
There is no evidence of physiological impairment in man at blood lead levels
up to 40 μg/100 ml that could be attributed to the progressive impairment
of this enzyme activity.

Lead also reduces the rate of transmissions of nerve impulses (Seppalainen, Tola, Hernberg and Kock, 1975; Hernberg, 1977). Signs of nerve
damage have been found in subjects whose blood lead levels, followed
throughout their working histories, had never exceeded 70 μg/100 ml,
commonly regarded as the safety norm for industrial exposure. The
relationship between levels of lead in blood and the rate of passage of
impulses in the brachial nerve appears to be direct and detectable down
to substantially lower concentrations, of below 30 μg/100 ml (Hernberg,
1977). The physiological significance of the changes detected at these very

low concentrations is, however, not clear and the neurological research done so far does not provide substantiation or explanation for some of the grosser symptoms (hypertension, aggressiveness and lowered mental acuity) which have been stated to correlate with quite low blood lead concentrations (Bryce Smith and Waldron, 1974; David, Clark and Voeller, 1972; Gibson, Lam, McCrae and Goldberg, 1967). Further research on these marginal physiological effects, and their biochemical basis, is clearly desirable, especially since exposure levels in some urban areas near to works using dusty forms of lead (DOE, 1974b) or in households receiving plumbosolvent waters through lead plumbing systems (DOE, 1977c) can be significantly increased.

In one analysis of such a population, living near a smelter, Lansdowne (1978) concluded that there was no correlation between blood lead levels and IQ or psychological disturbance and that lead levels were not a major factor affecting learning or behaviour: social influences, especially the stability of the home background were much more significant. Similarly, although some authors have suggested that the well-known correlation between softness of drinking water and the incidence of cardiovascular disease could be due to higher lead intakes because some soft (plumbosolvent) water dissolves lead from pipes and cisterns, there is no firm evidence that this lead is a cause of hypertension or kidney disease (Shaper, 1978). However, many people have also pointed out that lead is a substance known to have toxic properties, conferring no known benefits on the organisms whose bodies contain it, and commonly present in the human population at blood concentrations around half of those considered medically undesirable: the safety margin thus appears narrow and although detailed research may have failed to confirm certain possible hazards, it is clearly prudent to take reasonable steps to keep exposures down.

The relationship between low concentrations of lead in blood and ALA-D activity appears to be a case where a biochemical effect is 'contained' within the physiological regulatory capacity of the body, and does not spill over into other symptoms. The same may be true of low dosages of other metals. Many metals occur in enzymes and pigments that play a vital part in the chemical machinery of organisms. In excessive levels, elements such as lead, cadmium, mercury, zinc, copper or vanadium may act by substituting for the 'proper' metallic component, binding onto and interfering with the enzymes involved in synthesis, as lead does to ALA-D and other enzymes (Clayton, 1975). Rather different, but equally fundamental interference with biochemical mechanisms is caused by organochlorine pesticides. Of these, DDT and its derivatives or relatives such as DDD and DDE have been studied in especial detail and their effects or animals recently reviewed by Moriarty (1975b). DDT acts by interfering with the nervous system, through increasing the number of impulses that pass along an axon fibre in response to a stimulus. This is obviously a fundamental effect, one of whose consequences is to make insects and other animals poisoned by DDT hyperactive. Another commonly used type of insecticide, which is increasingly replacing DDT and its relatives, the organophosphorus group, interferes with the nervous system by inhibiting the enzyme acetylcholine esterase. This is an essential agent in removing acetylcholine which is released at a nerve ending when an impulse arrives there, and acts as a chemical transmitter allowing the impulse to pass the gap or synapse between adjacent nerve cells. If the acetylcholine is not removed it goes on triggering impulses, with inevitable disruptions in the co-ordination of the animal. Other enzymes whose function is less well understood are also likely to be affected by organophosphorus compounds.

DDT and organophosphorus pesticides thus have primary points of impact in the nervous system of the target organisms. But the lead and organophosphorus stories indicate another common feature: that a pollutant may have more than one biochemical primary point of impact. The same is true for DDT. This substance is well known not only to cause hyperactivity through its effects on the nervous system but to cause birds to lay thinner (and hence more breakable) eggshells than normal (Ratcliffe, 1970). Moriarty quotes a theory that DDT may affect the activity of the enzyme carbonic anhydrase which controls the combination of carbon dioxide and water in the tissues of the bird's shell gland to form the carbonate needed for the shell.

5.3.2 *Physiological responses*

Impacts like those described in the preceding section inevitably provoke gross physiological responses if the pollutant enters the body in any quantity, or exposure continues for any period. Again, the effects of DDT and other organochlorines have been examined most thoroughly in insects (the intended targets) as well as in mammals and birds.

Perhaps it is not surprising that DDT does not affect nerve impulse transmission in all parts of the insect nervous system at the same dosage level. Moriarty (1975b) reviews research in the United States by Soliman and Cutkomp which indicated that DDT at very low dosages (one-eighth of that needed to kill the most sensitive members of a housefly population, and one-quarter of that needed to produce visible symptoms of hyper-activity) increased the sensitivity of housefly sense organs to sugars in solution. Here the physiological response was not evidently deleterious – it led to the animals extending their proboscies to suck up liquid more readily than normal. It did indicate that it should, with careful study, be possible to work out the sequential symptoms of pollutant effect between the first physiological manifestations and the onset of death.

There are other examples of so-called 'sub-lethal' physiological effects of DDT and related substances. For example, it has been established that very low dosages of DDT or dieldrin can affect the behaviour and reproduction of fish and amphibians. Anderson (1971) demonstrated that salmon and trout from the rivers of New Brunswick showed changes in their cold resistance when exposed to as little as 40 parts per billion (10^9) of DDT. The lower lethal temperature was raised by 2 or 3 °C and this would be sufficient to cause higher mortality in winter, at the onset of migration. The DDT levels were comparable with those that can occur following aerial spraying to combat forest insects, so that the effect could well be a real one in nature. Rather similarly, Cooke (1970, 1975) has demonstrated that frog tadpoles become conspicuously overactive if kept for one day in a medium containing 5 ppb of DDT or 500 ppb dieldrin, and develop growth abnormalities if exposed to only 1 ppb DDT. These are DDT levels that could well occur in the field, and the hyperactive tadpoles have been shown to be taken by predators at above normal rate, probably because they are more conspicuous.

Low levels of DDT not only upset calcium metabolism in birds, leading to thin eggshells and probably raised mortality in the nest, but affect the thyroid and hence alter metabolic activity. Jefferies (1975) has carried out parallel research on the effects on birds of polychlorinated biphenyls – chlorine containing organic substances, but very different chemically from DDT. The thyroids of gulls and guillemots treated with PCBs at rates of 50 to 400 mg/(kg bodyweight day) are enlarged, developing goitres. Probably the direct effect is on the pituitary gland, causing a reduction in the output of the thyroid-stimulating hormone which regulates thyroid activity. The thyroids then enlarge, but diminish their output of thyroxine. The metabolic rate of the birds tends to decline. When the birds moult, they produce feathers with more widely spaced barbules, reducing waterproofing and heat insulation.

Cooke's work on frogs, and the eggshell – thinning studies illustrate another important principle. It is that in considering the effects of pollution on a species of animal (or plant, for that matter), the variation in susceptibility at different stages of the life cycle is of paramount importance. In tadpoles, sensitivity appears greatest in the first days after hatching. Mortality in a population then levels off (Figure 39) although the surviving tadpoles demonstrably contain pesticide residues. But at metamorphosis, probably because body reserves are mobilised and the residues contained there released, mortality again rises. The importance of stress and starvation in promoting such release of organochlorines from depot fat was described in Chapter 3.

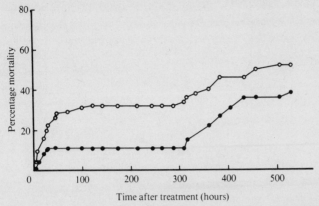

Figure 39. Cumulative mortality in groups of tadpoles treated with DDT. Solid circles, 1.00 ppm DDT group; open circles 10.00 ppm DDT group. The first phase of mortality soon after treatment is of tadpoles, and the second phase, after 300 hours, of froglets during metamorphosis. Control deaths were less than 3%. From Cooke (1970).

Changes in environmental temperature can also have substantial influence on the responses of cold-blooded animals because increased temperature raises metabolic activity. Moriarty (1975b) quotes his own research on grasshoppers, which showed that animals that had marked symptoms of DDT poisoning at 20 °C became apparently normal when transferred to 25 °C, only to display symptoms again when returned to the cooler environment. The consequence of all this is thus to demand surveillance of the exposure, stress and condition of an organism continuously if short periods of susceptibility are to be detected. Jefferies (1972) has recently shown this dramatically in bats, which need to feed intensively after emergence from hibernation and may acquire a lethal dose of insecticide residue in a fortnight or so at this stage, whereas they would be in no risk from the same level of prey contamination when feeding more lightly throughout the rest of the spring and summer. In searching for subtle physiological effects, therefore, the nutritional state, life cycle stage, and environmental influences upon an animal all need to be taken into account, and if toxicity tests are done the results may be expected to vary with such factors (as Figure 40 illustrates). The age, physiological state and environmental conditions are, indeed important in man under real life circumstances: various authors have, for example, pointed out that human susceptibility to lead is highest in children (see DOE, 1974b).

The effects of low levels of a pollutant on biochemical and physiological systems need not be wholly adverse. Many environmental contaminants, including PCBs and organochlorine pesticides, stimulate the liver to synthesise more of the enzymes that break the contaminants down. This

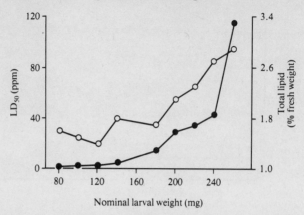

Figure 40. Graph for fifth instar caterpillars of the small tortoiseshell butterfly to show how the toxicity of dieldrin and the percentage of body fat change with caterpillar weight. Open circles, concentrations of total fats; solid circles, LD_{50} for dieldrin. From Moriarty (1975b).

is an obvious defensive mechanism and helps to explain why some biochemical effects are 'contained' by the homeostatic machinery of the body. Unfortunately, the enzymes involved are not always specific to the contaminant and may also break down steroids, including sex hormones: even this need not, however, cause serious upsets to reproductive behaviour since the body's capacity to increase the production of these substances may be triggered. The critical questions concern the scales, rates and interactions of such processes.

5.3.3 *Carcinogenesis, mutagenesis and teratogenesis*

The action of radiation in altering DNA and hence producing certain cancers was described in Section 5.2.1. Chemical agents are also capable of altering DNA structure or of interfering with the biochemical processes involved in its replication and in the development of organic structures which it controls. Other chemicals can interfere with the biochemical mechanisms that allow organisms to repair such damage or mitigate its effects. Carcinogens are very diverse in structure, but appear alike in being changed by metabolism within the target into strongly reactive molecules that in turn interact with nucleic acids and other large 'macromolecules' (Garner, 1978). Besides cancers, mutations and embryonic abnormalities can result from these biochemical impacts.

Only a minor part of the mutation found in animals results from cosmic and other natural radiation. Even when the increment due to man's release of radioactive materials and pollutants is added, the total may account for only 5 % or so of all mutations, most of which are due to biochemical imperfections and instabilities (Vogel, 1978). None the less, it is known that the reproductive processes in man and other animals may be seriously affected by environmental chemicals. The scale of this effect is hard to judge because mutations, whether natural or induced, generally so impair the development of embryos that they die and are aborted, only 2% or so surviving to be born (Vogel, 1978; Sullivan, 1978). This spontaneous abortion is, of course, a valuable biological defence mechanism but the increase in miscarriage or infertility caused by exposure to chemicals in the environment (especially the working environment) is obviously unacceptable.

Such effects have been known for a long time. In the early years of the present century it was established that the rates of miscarriage and neonatal mortality were both unusually high in women working in the lead industry: 11% and 40% respectively in one study cited by Sullivan (1978). Premature births were also abnormally common, and lead levels were found to be elevated in the placentas of malformed foetuses. Male infertility was also found to be double the normal in some groups of lead workers. Lead is not the only substance capable of such effects. Organochlorine pesticides are reported to suppress ovulation in women. Methyl mercury, like lead, traverses the placenta and has a higher affinity for foetal than adult tissues, and in the Minamata disaster deaf, blind or spastic children were born to mothers who themselves appeared unaffected. Sometimes, however, susceptibility may be greater the other way round and Sullivan (1978) cites one illustration: 63 out of a group of 95 women who died of beryllium poisoning after exposure at work were pregnant. In this case, susceptibility to the toxic chemical was presumably increased by the stress caused by pregnancy. Miscarriages among women working in certain hospitals are also known to be well above normal, perhaps in some cases due to anaesthetics and in others to dioxins contaminating hexachlorophene used as a disinfectant. These and other examples are set out in detail in the Report of the Royal Society Discussion by Doll and others (1978).

5.3.4 *Effects on animal populations*

The discovery that PCBs in birds can affect metabolic rate and also the structure and waterproofing properties of the feathers produced at the annual moult may help to explain the sequence of events leading to one of the most spectacular catastrophes among wild birds in recent years – that in the Irish Sea in the autumn of 1969 when between 12 000 and 15 000 seabirds died (Holdgate, 1969a, b). Guillemots (*Uria aalge*) were chiefly affected, and the victims were almost all adults in the period just after the annual moult, which in turn follows breeding. Birds began to come ashore in northern Ireland during fine weather in late August and early

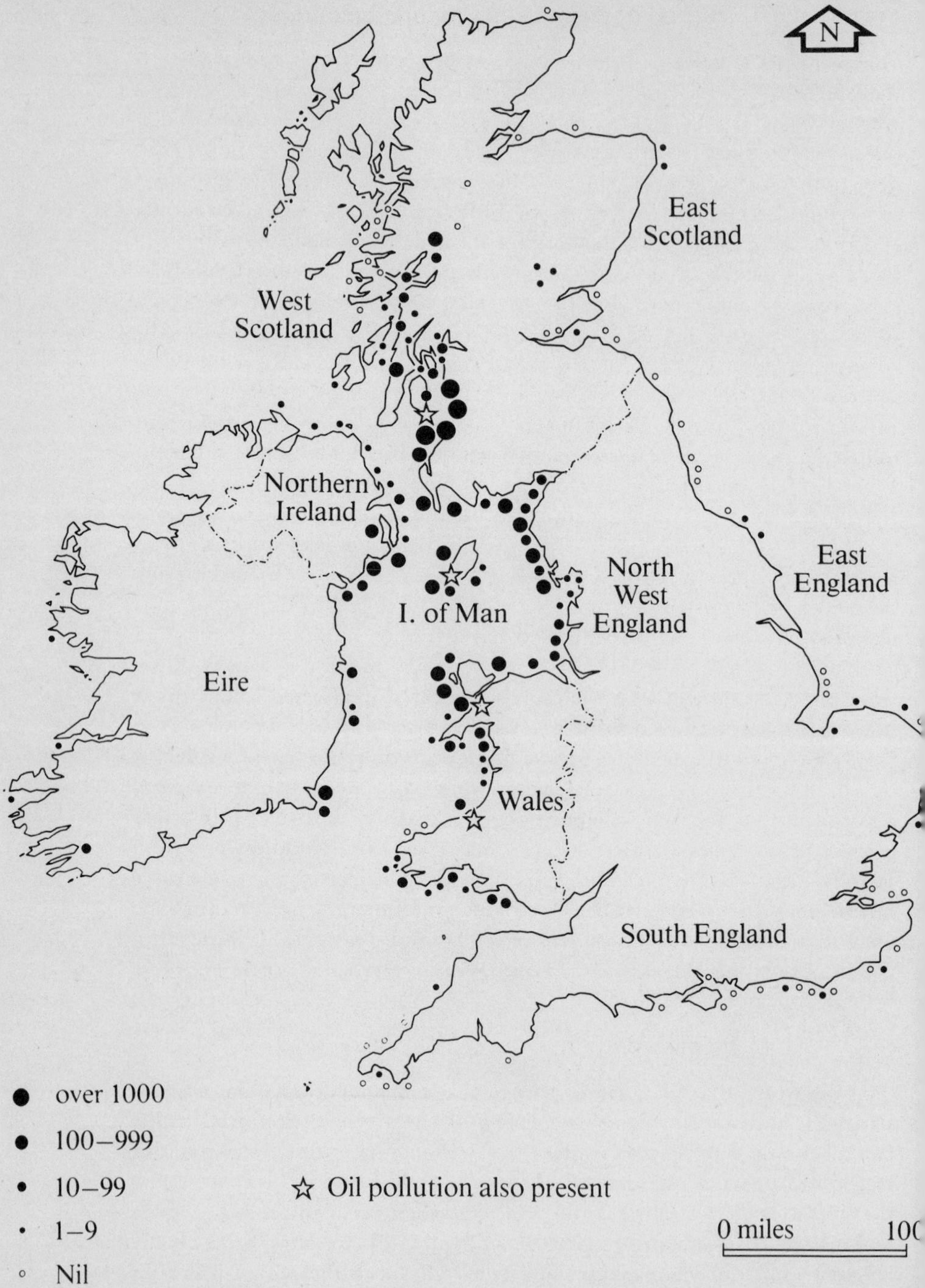

Figure 41. Maximum counts of dead Auks, September–November, 1969. From Holdgate (1969b).

East
Scotland

West
Scotland

Northern
Ireland

East
England

North
West
England

I. of Man

Eire

Wales

South England

● over 1000

● 100−999

· 10−99   ☆ Oil pollution also present

· 1−9

◦ Nil

0 miles   100

Table 16. *PCB levels in the Irish Sea and in sea birds, during and after October 1969*

Sample	PCB concentration (μg/kg wet weight)
Sea water	< 0.01
Zooplankton	< 10–30
Mussels	10–800
Herring	
muscle	< 10–2600
liver	10–2000
Whiting	
muscle	10–400
liver	4500–27000
Guillemots[a] (found dead)	
liver	2000–880000
Guillemots (found dead)	
liver	10000–200000
rest of body	1000–10000
Guillemots (shot, apparently healthy)	
liver	0–2000
rest of body	1000–7000

[a]   *Note*: the first entry for guillemots found dead is for the main sample of 49 birds examined. The second refers to a random sample of 5 birds chosen to compare with a sample of 5 apparently healthy birds. It will be noted that the PCB concentrations in the body tissues of the shot birds and those found dead are comparable, but the liver concentrations are much higher in those found dead.

September. They were thin, though not emaciated, and described as swimming low in the water (possibly they were partly water-logged). When storms blew up in September the main wreck of exhausted and semi-emaciated birds followed, especially on the coast of southern Scotland (Figure 41). Examination showed that PCB levels in the livers of the dead birds were high (Table 16).

Healthy birds shot from an area further north where there was no wreck despite even more violent storms, also proved to have a high body burden of PCBs, but this was present largely in the body fat and later research suggests that on average the levels were only about half those prevailing in the Irish Sea. The hypothesis that best accounts for the chain of events needs to start by accepting this slightly higher loading with pollutants in the area where the deaths occurred. The levels may have been high enough to cause some impairment of feather replacement in the moult, and some reduction in metabolic rate. This could have led to a reduced rate of feeding success in the birds, which would respond by mobilising body fat, releasing PCBs into the blood and aggravating the situation. The further stress of autumnal gales would inevitably have made matters much worse, by reducing feeding rates, increasing body fat mobilisation and making any reduced efficiency in waterproofing and insulation an added liability. Hence the rapid and catastrophic build up of mortality. Even if this is not the complete story, the point is an important one. Accumulation of a pollutant, perhaps with only minor initial effects, seems to have spilled over into major physiological impairment which, under environmental stress, resulted in an effect at the population level. The effect may be to some extent analogous with the increased sensitivity to beryllium in pregnant women mentioned in Section 5.2.4.

The population changes in herons and predatory birds in the 1960s demonstrate a comparable effect, but mediated at a different stage in the life cycle. Herons suffered considerably from organochlorines at the peak of utilisation of these pesticides in Britain. There were deaths among adults through acute poisoning. There was also a production of many thin-shelled eggs, with consequent mortality in the nest. Yet these effects did not spill over to the population level in this species. The reason was that despite the reduced efficiency and raised adult mortality, enough young birds were still being produced for recruitment into the breeding population to be sustained. What did happen was that the average age of the breeding population dropped. On the other hand in birds of prey such as eagles and peregrine falcons the combined effects of acute mortality among adults (especially due to dieldrin) and mortality in the nest due to breakage of thin-shelled eggs appear to have combined to lower numbers dramatically. Other factors (myxomatosis reducing rabbit populations, habitat changes and human persecution) may have contributed, but it is generally held that organochlorines were the dominant cause of this population effect. But since these predators, at least in the man-dominated ecosystems of Britain, are not very important determinants of ecological characteristics, there was no evident response at the ecosystem level (although no doubt there was a change in the relative importance of different causes of mortality among prey species, and maybe some marginal changes in population sizes).

These effects are a reflection of the operation of basic mechanisms of population limitation in animals. There are four vital parameters in these mechanisms:

(a) numbers of young produced per annum;
(b) numbers of young dying before recruitment into the breeding population;
(c) number recruited into the breeding population per annum;
(d) number of breeding adults dying per annum.

It is an over-simplification to say simply '$c = d$' and '$b = a - c$' because in most species the breeding adults survive for many years, recruits are drawn from individuals several years old, and the mortality rates all vary annually. None the less the crude relationship is valid. The point about it, in the present context, is that pollutants can operate at several points on the chain. They can increase deaths among the breeding adults. They can reduce the production of young through sub-lethal effects on breeding performance (like production of thin-shelled or sterile eggs in birds, or stress on adult mammals causing embryonic resorption *in utero* or abortion). They can increase juvenile mortality, slimming the numbers available for recruitment. Populations have a finite capacity to compensate for a certain level of mortality because of the production of a surplus of young, normally eliminated by many factors including competition with the established adults. If adult numbers are reduced, that competition is relaxed and more young survive to recruitment. In this we have an exact parallel with the situation when a wild population is cropped by man, discussed in Chapter 1, and as in that situation, it is when pollution stresses a population beyond the limits of its reproductive resilience that a decline or even a crash ensues.

So far, we have only considered cases where pollution acts directly to reduce a population. But there are also indirect effects. Reduction in a prey species may cause its predators to change both in numbers and diet, thereby affecting populations of yet other species: thus brook and rainbow trout in the streams draining forests in Idaho became heavy predators of crayfish when aerial spraying of DDT eliminated many of their normal food organisms (Dempster, 1975). Similarly, insecticidal sprays are well known to be capable of causing increases in some resistant species (like the red spider mite, *Panonychus ulmi*) because the many predators of this naturally rare organism are more sensitive than it is, while changes in the composition of soil fauna in sprayed areas may result from higher mortality among predatory mites than among prey Collembola. These are all examples of the ecological side-effects of actual control spraying rather than inadvertant contamination of vegetation or soil with persistent organochlorines, but the boundary is a narrow one and it is likely that sub-lethal effects of residues of such substances will have ecological consequences away from the site, or well after the time, of deliberate application (see Dempster, 1975, for illustrations of sub-lethal effects of organochlorine applications).

5.4 Effects on plants

The effects of pollutants on plants range in a similar fashion from observed changes at the biochemical and cellular level through overall effects on growth and performance to major alterations in population size and species distribution (see Holdgate, 1979 for a recent review).

5.4.1 *The nature of the effects*

Gaseous pollutants such as sulphur dioxide, ozone, or nitrogen oxides enter plant leaves especially through the stomata and affect the palisade and mesophyll cells of the leaf, causing cell collapse, damage to the plastids, and hence irreversible loss of photosynthetic capacity (ARC, 1967; Saunders, 1976; Holdgate, 1979). Fluorine and sulphur dioxide affect lichens in an analogous manner, killing the algal cells in the symbiotic partnership on which the plant as a whole depends for photosynthesis. The precise biochemical mechanism of these impacts is however uncertain.

5.4.2 *Interaction between pollutants, and between pollutants and other variables*

Although it is short in terms of distance, the final section of the pathway by which a pollutant reaches a plant is of critical importance. Gaseous pollutants move at a rate which depends on the diffusion resistance over the leaf, and at the stomata, and this is greatly affected by the rate of airflow over the leaf, the roughness of the surface, and whether the stomata are open or shut (Ashenden and Mansfield, 1976). Whether stomata are open is in turn influenced by the physiological condition of the plant (and especially its water balance), by atmospheric relative humidity, and by sulphur dioxide concentrations: above a relative humidity of around 50–60% moderate sulphur dioxide concentrations promote stomatal opening. Particulate pollutants may influence the process by blocking stomata, reducing their aperture but also preventing total closure. In contrast, liquid droplets of pollutant, including aqueous droplets with pollutants in solution, may depend for their effect on the composition, texture and degree of contamination of leaf and stem surfaces while particulates may be of greatest significance because they abrade surfaces or disrupt protective wax layers by adsorption. Other pollutant pathways to plants via the soil and soil water depend critically on how readily the substances pass through the root surface or are translocated from root to shoot, and in many plants there appears to be a barrier to the passage of metals such as lead.

These variables probably account for considerable differences in the experimental measurements of damage obtained by different investigators even for relatively well-studied interactions like that between sulphur dioxide and sensitive target plants such as the ryegrass S23. Generally speaking this grass appears to be more sensitive in winter than in summer, and to show impaired growth at concentrations of sulphur dioxide above

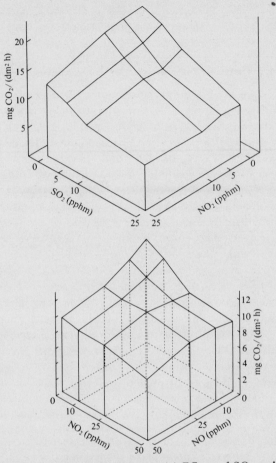

Figure 42. Interactive effects of pollutants. (*a*) Effects of SO_2 and NO_2 pollution
on the rate of net photosynthesis in pea, *Pisum sativum*. From Bull and Mansfield,
1974. (*b*) Effects of NO and NO_2 pollution on the rate of net photosynthesis in
tomato. From Capron and Mansfield, 1977. Note that in both cases combinations
of pollutants have greater effect than either gas does alone.

about 50–100 $\mu g/m^3$ (Bell and Mudd, 1975). However, sulphur dioxide,
although the most universally monitored of potentially damaging air
pollutants, is by no means the only phytotoxic gas to occur in the air of
industrial regions and it is now clear that such gases may have greater effects
acting in combination than alone. Sulphur dioxide, oxides of nitrogen
and ozone in various combinations appear likely to enhance one another's
impacts (Mansfield, 1974; Capron and Mansfield, 1975; Bull and Mansfield,
1976) whereas ammonia and sulphur dioxide may neutralise each other's
effects (Saunders, 1976) – though the precise situation will undoubtedly

vary from gas mixture to gas mixture, and according to the influence of other variables (Figure 42).

5.4.3 Variation between species and taxonomic groups

Genetic factors are of great importance. Some plant groups and species are well known to be much more sensitive to pollutants than others. Lichens are the most sensitive of known plants, and their growth is progressively impaired at mean atmospheric sulphur dioxide concentrations above 50 μg/m^3 (Ferry, Baddeley and Hawkesworth, 1973). Some bryophytes are affected by sulphur dioxide at above 60 μg/m^3, some fungi at 100–150 μg/m^3 and coniferous trees in the range of 100–180 μg SO$_2$/m^3. One obvious consequence of this differential sensitivity is the absence of ornamental conifers from the centres of many towns, and the presence instead of a restricted range of hardy trees like London plane (*Platanus* × *hispanica*).

Within species there are similar variations. Tobacco strains differ considerably in their sensitivity to ozone (Bell, 1974). The ryegrass S23 is relatively much more sensitive than strains adapted to polluted regions (Bell and Mudd, 1975). The growth of ryegrass from Helmshore in the south Pennines, for example, is better under local conditions than that from a less-polluted coastal locality but the relative performances are reversed when the habitats are transposed. This pattern, clearly an indication of the evolution of local races adapted to differing levels of atmospheric pollution, parallels the well-known evolution of metal tolerant strains on soils polluted by former mine workings (Wainwright and Woolhouse, 1975).

5.4.4 Patterns of plant performance in nature

Plant growth is the outcome of many factors, among which pollutants must clearly be counted. There are numerous records of plant performance which can be related to pollution, and some examples come from the aquatic as well as the terrestrial environment. Burrows (1971) demonstrated that the growth of the seaweeds *Ulva lactuca* and *Laminaria* varied from one kind of sea water to another in a fashion that appeared correlated with pollution (Figure 43). It has accordingly been suggested that the varying growth rates of these algae could provide a useful biological indicator of the cleanliness of the marine environment. Lichens, the impairment of whose growth by sulphur dioxide has already been noted, are being used in this way (as described below). Over wide areas of the United States concentrations of ozone and other oxidants produced by photochemical reactions involving vehicle emissions are sufficient to check the growth of sensitive spinach, lettuce and tobacco strains and locally harm conifers

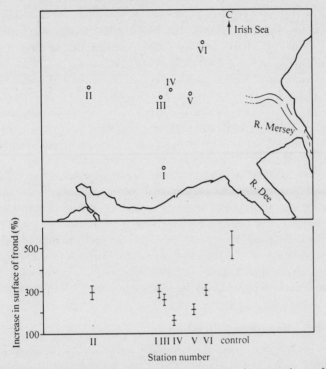

Figure 43. Growth rate, measured as percentage increase in surface area, of *Laminaria saccharina* over a 3-week period in culture media made up with waters from different sampling stations in Liverpool Bay. Station III is the northwest Light Buoy, near which sewage sludge and industrial waste have regularly been dumped. Visible evidence of pollution was noted at Stations III, IV and V. From Burrows (1971).

(Bell, 1974), and this 'oxidant smog' checks tree growth within 100 km of Los Angeles and is said to make citrus growing impossible within 50 km of that city (Figure 25). Losses to agriculture through the inability to grow productive grasses like S23 ryegrass in polluted zones such as south-east Lancashire have been calculated over many years (Bleasdale, 1959).

These are all examples of pollution effects on plants at the population level, in some cases excluding species or varieties from otherwise suitable habitats. The phenomenon can occur at many scales, and may involve mixtures of pollutant as well as single substances. For example, lichen floras are impoverished around aluminium smelters and the species that persist have enhanced levels of fluorine in their tissues (Perkins, 1975), but such smelters may also emit sulphur dioxide, hydrochloric acid, and particulate aluminium and fluoride so that it is not always clear which was the cause of the observed effects. It is even more difficult to be sure about interactions involving long pathways and complex transformations, like those involved in the deposition of sulphur dioxide, oxidised as sulphate in 'acid rain' over regions such as southern Scandinavia. While such transport and deposition undoubtedly occurs, the interplay of local and distant sources, and of 'strong acids' deposited in rain and 'weak acids' like humid and fulvic acids formed locally in soil and litter, and the effects of both on plant growth and forest productivity is proving hard to unravel (DOE, 1976c; ITE, 1976). Under such circumstances uncertainty may be increased because of indirect effects including changes in soil conditions and differential effects on competing species or on parasites and hosts. Over parts of the English midlands and Scottish borders moderate air pollution with sulphur dioxide may be beneficial to some crop plants because it remedies a natural shortage of sulphur in the soil, while in many urban zones rose mildew and rose black spot diseases are eliminated by sulphur dioxide levels that appear to have no effect on the performance of the host plant. Such changes in the structure and composition of exposed plant communities are hard to unravel and complicate an analysis of all but the most obvious effects of pollution at the ecosystem level (Williams and Ricks, 1975).

5.5 *The ecosystem level*

Ultimately, changes in plant and animal populations must combine to change whole ecological systems. There are several well-documented examples of the effects of pollutants upon plant communities and their associated animals. One of the best examples concerns communities of corticolous lichens – that is, those growing on tree trunks (Ferry, Baddely and Hawkesworth, 1973). This is a habitat that is conveniently standardised for observation, especially in Britain where oak trees planted well-spaced from one another in the parks of the eighteenth century can be found all over the country. Hawkesworth and Rose (1970) have shown a progressive impoverishment of this corticolous lichen community that can be correlated with exposure to air pollution (Figure 44) especially by sulphur dioxide. Ten zones of increasing impoverishment have been recognised. It seems clear that over the central midland plain of Britain there has been a major reduction in lichen abundance and diversity. Progressive impoverishment, and species change, as air pollution increases, must genuinely affect a whole ecosystem, for herbivorous insects (Psocidae) that feed upon the lichens are likely to decline in parallel with them, in turn affecting predators.

There is evidence of ecological changes in south Pennine peat bogs over the years since the Industrial Revolution. Conway (1943) demonstrated at Ringinglow bog, near Sheffield, a transition from sphagnum mosses to cotton sedge (*Eriophorum vaginatum*) around the year 1850. This could well be one of the effects of the massive increase in air pollution accompanying

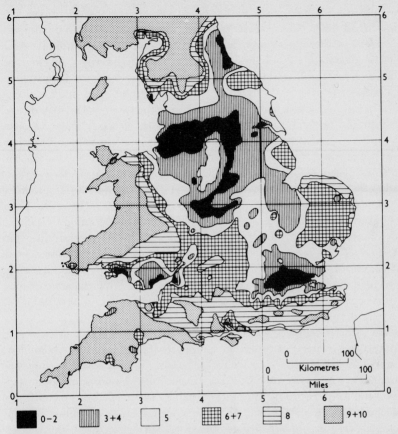

| ■ 0–2 | ▨ 3+4 | □ 5 | ▦ 6+7 | ▤ 8 | ▨ 9+10 |

Figure 44. Approximate boundaries of zones of richness of lichen floras in England and Wales. Zones 9 and 10 are richest in flora and Zones 0–2 poorest. The correspondence with the pattern of urbanisation is evident. From Hawksworth & Rose *Lichens as pollution monitors* (Studies in Biology No. 66) E. Arnold.

the growth of the industrial towns nearby. Acids (sulphuric and nitric) produced by the solution of sulphur trioxide and oxides of nitrogen in rain, and 'washed out' into lakes may also cause the pH of the water to fall to as little as 4.0 and at this level of acidity many fish (including trout) cannot survive and the structure of fresh-water systems may change dramatically (DOE, 1976c; Sweden, 1971). Dramatic changes in fresh waters also occur, on a far wider scale, as a result of 'eutrophication' (OECD, 1970). (Literally 'eutrophic' means 'well-fed' – but not necessarily excessively so. The phenomenon of over-nutrition to which the term is now commonly applied should really be named 'hypertrophication' or 'over-feeding'.) The process is of interest in demonstrating the harmful consequences of a basically beneficial action: the discharge of nutrients that stimulate plant growth. Nitrate and phosphate are the principle nutrients, derived from sewage treatment, detergent degradation and from fertilisers and from the enhancement of natural processes of nitrification in soil through more efficient husbandry. When land is ploughed, nitrate produced by bacteria is inevitably released, and the more efficient the farming practice the higher the levels of this natural nitrification are likely to be. Passing into rivers and streams, these nutrients promote the growth of water plants which are often limited by a natural shortage of nutrients (Lee and Veith, 1971). The 'bloom' may be of many species: in lakes small floating single-celled diatoms are most common while in rivers and streams, filamentous green algae (e.g. *Cladophora*) and rooted plants all respond well. In some lakes, the bloom of planktonic algae may be so dense that it shades out and kills the plants on the lake bed.

The problem with this excessive growth is that it commonly develops faster than the herbivorous animals in the ecosystem can multiply and consume it. The rapid growth is therefore followed by the production of quantities of dead plant matter which are degraded by the bacteria, using up oxygen in the process. So much may be consumed that the green plants in the ecosystem cannot renew it fast enough by photosynthesis. Concentrations of dissolved oxygen may then fall too low for the water to sustain certain animals (salmonid fish such as trout and salmon are particularly sensitive), or even to the point where there is no oxygen at all and the whole ecosystem changes, aerobic species disappearing and being replaced by anaerobic bacteria, small tubificid worms, certain insect larvae and a few other invertebrates capable of living in such conditions. Some authors describe such an anaerobic system as a 'sulphuretum' because of its emission of hydrogen sulphide. Such anaerobic communities are less well able to decompose further organic matter and if raw sewage or more dead plant material is added, it may persist for a long time. The result is a great loss of amenity, a possible health hazard, and through the emission of hydrogen sulphide, an unpleasant smell.

Another pollutant capable of widespread ecological effects is oil. It is commonly stated that oil is mainly significant because of its physical effects (Chapter 4), which are especially manifest in birds whose buoyancy is destroyed when their plumage is fouled (although this effect is aggravated when their natural preening behaviour leads them to swallow toxic oil constituents). But 'oil' is a complex mixture of hydrocarbons, varying widely in chain length and structure (Baker, 1976a; GESAMP, 1977). Some North Sea oils, for example, contain significant amounts of short-chain alkanes and aromatics. These are relatively soluble in water and are more toxic than the longer-chain compounds (DOE, 1976d). They might, therefore, affect sensitive floating organisms (plankton), for if absorbed (and they are also fat-soluble) they could well affect the enzyme systems that control fat and steroid metabolism. Such effects have not yet been established: nor has there been proof of a link between the presence in the sea of potential carcinogens in the oil and the development of carcinomas in marine life or human consumers of tainted seafood. There is no evidence, either, of the accumulation of hydrocarbons derived from oil in the tissues of fish or shellfish in the North Sea, but research in all these fields is continuing – rightly in view of the increasing exploration and exploitation of undersea oilfields. Another important area of study is the rate and process of breakdown of oil at sea, which can be rapid but depends on many factors including the composition of the oil, the temperature and roughness of the sea, and the abundance of marine bacteria.

The effects of oil on shore ecosystems have been studied much more fully than the biochemical and physiological effects of oil residues and breakdown products. Repeated pollution around refinery outfalls in Milford Haven and at Fawley led to the reduction in the biomass and diversity of the marine fauna and salt-marsh flora (Baker, 1976b; Dicks, 1976; Regier and Cowell, 1972). The treatment of shores with detergents to disperse the oil has often had a more dramatic effect than the oil itself, especially since the detergents used at the time of the *Torrey Canyon* wreck were not only more toxic than the oil but up to 100 times more toxic than those now in use. Shores sprayed in 1966 were often cleared of life (Smith, 1970) and when re-colonised, green algae grew unusually vigorously because the limpets which normally keep them in check by grazing recolonise the rocky surfaces more slowly. It took several seasons for a 'normal' ecosystem to re-form.

Oil is an example of a pollutant whose ecological effects vary widely according to the precise composition of the oil and the circumstances of exposure. The authoritative review by GESAMP (1977), backed by a comprehensive bibliography, leads to the conclusion that in the open sea both single and repeated oil spills have so far posed greatest hazard to seabirds but that light and refined oils can also hazard planktonic eggs and larvae of fish. Life in the intertidal zone can be severely affected and since recovery takes several years, repeated oilings can have a profound effect on the structure of ecosystems. Coral reefs present especial problems, for if the coral organisms are killed, natural protection from erosion may be lost, and since many coral islands are very isolated their recolonisation by organisms formerly present in the ecosystem may take a long time.

There are two general features of damage to ecosystems by pollution: species diversity is progressively reduced and the productivity of the ecosystem is impaired. The basic conceptual model of such a sequence, and some supporting evidence from salt-marsh vegetation, has been provided by Regier and Cowell (1972). They have demonstrated in Milford

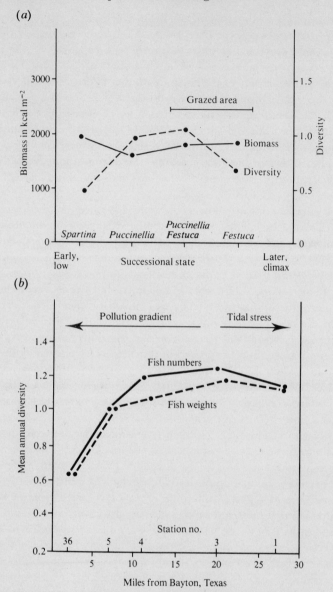

(a)

(b)

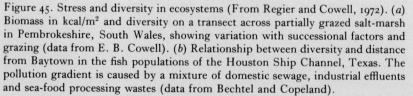

Figure 45. Stress and diversity in ecosystems (From Regier and Cowell, 1972). (*a*) Biomass in kcal/m² and diversity on a transect across partially grazed salt-marsh in Pembrokeshire, South Wales, showing variation with successional factors and grazing (data from E. B. Cowell). (*b*) Relationship between diversity and distance from Baytown in the fish populations of the Houston Ship Channel, Texas. The pollution gradient is caused by a mixture of domestic sewage, industrial effluents and sea-food processing wastes (data from Bechtel and Copeland).

Haven salt marshes that the species diversity can be inversely correlated with grazing stress, and review some evidence that pollution reduces fish species diversity and biomass (Figure 45). But they point out that the initial effect of pollution may be to increase diversity, the decline only ensuing as the level rises further. In terms of the thesis developed in Chapter 1, this phenomenon may occur because the stress of pollution initially opens up parts of the ecosystem by eliminating some individuals – K-strategists being particularly vulnerable. New colonists can then enter the system. But mounting stress eliminates more species than can exploit the opened habitat, and in the latest stages erosion may add further physical stresses, accelerating the process of degradation.

Figure 34 (which is a basic diagram of some of the principal elements in an ecosystem) illustrates the many points at which the processes in such systems can be affected by pollution. The various sections in this chapter have illustrated how individual pollutants act by perturbing biochemical processes: how these impacts in turn affect physiological performance, reproductive success and the dynamics of populations, and how, finally, changes in the performance of populations of individual species may interact to modify ecosystems. But (as Chapter 1 emphasised), pollutants cannot be considered singly and in isolation from other variables. Pollutants may interact enhancing or reducing the severity of one another's impact, and other stresses may likewise enhance or impair the tolerance of targets. We are forced to deal with a great complex of variables, and within this complex to treat pollutants as some among many chemical factors that affect the structure and functioning of living systems. Mathematical models that have been developed to describe the operation of ecosystems can equally be adapted to describe their perturbation by pollution, if sufficient is known about the component processes and their sensitivities. It is through the further development of such models that we are most likely to gain tools to explore how ecological systems are likely to respond to increases in pollution of various kinds – or to the measures we adopt in order to abate pollution.

6. Target exposure, risk and the establishment of goals and standards

6.1 *Risk estimation*

Pollution is controlled in order to protect targets from unacceptable damage. The effects of pollutants are studied in order to determine what levels of exposure, over what periods of time, in the presence of what other environmental variables will be liable to produce a particular amount of damage. As Section 4.1 explained, the product of exposure and time gives the dose to which a target is committed under particular circumstances, and if criteria relating dose to damage have been thoroughly established, harm commitment can be calculated and its acceptability judged. This has been done especially carefully in the radiation field (from which the terms dose commitment and harm commitment originate) (UNSCEAR, 1972; ICRP, 1966, 1973; SCOPE, 1977). Section 4.3.4 illustrated how these concepts could also be applied to the possible attenuation of stratospheric ozone by various pollutants, with consequent increases in the dose of UV-B people would be likely to receive, and Section 5.2.1 discussed the possible resulting damage from skin cancers.

Sometimes, in such an analysis attempts are made to calculate *thresholds* – that is, minimum doses liable to cause the maximum amount of damage considered tolerable. Threshold Limit Values (TLVs) are set by some agencies, defining the highest dose it is considered safe for somebody (usually a worker in a particular industry) to sustain. The drawback to such calculations is that they commonly relate to a 'standard' man – for example an adult male weighing 150 lb – and take little account of target variability. The concept of a standard or 'reference' man (or 'reference' cow for that matter) has its uses, but it is important that the figure for average human or bovine response is flanked by the statement of the standard deviation. Threshold limit values for adult male workers may, for example, be of little use in assessing the likelihood of damage to pregnant women or to children, whose special sensitivities to some substances were mentioned in Chapter 5.

Figure 46 emphasises this by pointing out that at (say) 60% of the dose needed to kill all the target population a few may die, 60% show major physiological disorders, and 35% or so display only minor physiological or behavioural change. These variations in susceptibility may well reflect

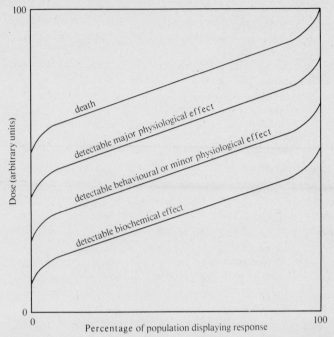

Figure 46. Schematic pattern of variation in response to pollution within a population. The lines represent threshold doses at which the various responses are first detectable.

age differences, genetic variation or differences in past exposure. The figure illustrates a purely hypothetical case, but the pattern is valid and it affects the steps taken to provide protection.

Because every member of a population of targets does not respond to a particular level of a pollutant in the same way, unless every member is assessed individually predictions of the consequences of such exposure can only be made in statistical terms – that is, in terms of the probability of a certain proportion of deaths or physiological upsets in a particular period. This need to evaluate likely effects in terms of probabilities is increased by the variability in the behaviour of pollutants in the environment, in response to shifts in weather or air and water flow. Such an attempt to specify the probability of a particular level of hazard is termed risk estimation and it is adopted in many contexts outside the pollution field (Table 17). The risk of major hazard through the failure of industrial plant (especially vessels containing toxic materials) for example needs to be calculated and offset against the benefits to the community of the process involved. The risks of accident associated with different transport systems are widely debated, and this has often led to action to reduce them by

Table 17. *Some voluntary and involuntary risks*

	Risk of death/person/year ($\times 10^{-5}$)
Voluntary activities	
Smoking (20 cigarettes/day)	500
Drinking (1 bottle wine/day)	75
Football	4
Car driving	14
Motor cycling	2000
Involuntary activities	
Influenza	20
Leukaemia	8
Run over on the road (UK)	6
Floods (USA)	0.22
Earthquake (California)	0.17
Lightning (UK)	0.001
Falling aircraft (USA)	0.001
Transport of fuel or chemicals (USA)	0.0005

Figures from Kletz (1977): see also Pochin, 1975.

building safer cars or better highways or making crash helmets or seat belts or air bags compulsory. Discussions of such processes of risk estimation are to be found in many official reports (Committee of Major Hazards, 1976; SCOPE, 1976).

The assessment of risk from pollution should be a strictly scientific process: a matter of evaluating probabilities using the best available information about the dispersion of a pollutant and the responses of targets. Once the likelihood of a particular level of effect under particular circumstances of emission and dispersal has been determined, value judgements enter into the picture: we are in the later stages of the basic model set out in Figure 33 and are dealing with effects on the socio-economic system. This is especially the case if there is a no threshold situation in which *some* undesired effect inevitably follows any exposure of a target. The socio-economic system responds by determining just what kinds of effect (how much harm commitment) is acceptable, and then taking measures to control the pollution so that (assuming the risk estimation was well done) adequate protection is assured. The actual controls involve limitations on the emissions of potential pollutants to the environment or the exposure of targets in one way or another, and they may be associated with set standards or objectives.

Such an approach is not without its critics. Criticisms stem in part from the inadequacy of the data on which risk assessments are based. For example, the uncertainties concerning the effects of chronic exposure mentioned in Chapter 5 naturally make it almost impossible to quantify the associated risks. Interactions are not always examined: for example calculations of fatal-accident frequency in industry are often done on a sectoral basis, failing to take account of mortality caused by the products of one industry to the workers in another area. Further criticisms stem from the concept of risk acceptability – acceptable to whom? Finally, it is a fact that disparate risks are encountered and disparate efforts made to reduce them. There are good reasons (e.g. in papers by Peto, 1978 and Doll *et al.*, 1978) for believing that risks of acute, and probably even of chronic damage, from general environmental pollution are much less than those from exposure to chemicals in the place of work, even in the most polluted urban zones. These risks are much less than those due to smoking, drinking, and accidents while travelling or at home: does this mean that pollution control should receive less emphasis than it at present does? Different industries devote widely differing efforts to protect workers from hazard: Kletz (1977) cites an analysis by Craig Sinclair showing that in agriculture in Britain £2000 is spent to protect an employee's life, while in steel handling the figure is £200000 and in the pharmaceutical industry, £5 million.

Faced with such uncertainties and imbalances, some (e.g. McGinty and Atherly, 1977) suggest that it is more logical to take all practicable steps to eliminate avoidable risk irrespective of the niceties of precise quantification. However, the prevailing approach to pollution control is based on the principle that, with scarce resources, attempts should be made to relate the control imposed to the hazards of exposure, and this approach will be followed in the remainder of this chapter. The compromise between the two approaches may lie, in any event, not in abandoning an attempt at the judgement of risk but in setting safety margins that themselves take account of the degree of uncertainty.

6.2 *Standards and objectives*

Standards are statements about the levels of target exposure or pollutant concentration that are considered acceptable at particular times and under particular circumstances. As SCOPE (1977) points out, there are two main types:

(a) those specifying *maximum* permissible levels of environmental contaminants (or target exposures or doses);

(b) those specifying *minimum* permissible qualities of performance of a human product (e.g. factory ventilation or dust arrestment: this kind of standard is applied to many types of building, like dams and bridges, outside the pollution field).

A distinction is often drawn between *standards* and *goals* or *objectives*. Both may specify desirable maximal target exposures or pollutant concentrations, but goals and objectives are not normally backed by law, and may not yet be attainable (at least at a socially acceptable cost): they are statements of policy aims. Standards, on the other hand, are formulated as a part of regulatory operations and have at least some basis in law. The process of devising both is, in scientific terms, closely similar: the distinction depends on the administrative context.

6.2.1 *Primary protection standards*

Standards of the first type can be defined for any point on the pathway from source to target (Figure 47). The target, however, provides the most logical starting point. This is why the most fundamental standard is the *primary protection standard*, defined as the maximal allowable exposure (or dose) for a particular type of target. Even here, however, there are complications.

Such standards can be set at the 'surface' of the target (as in Figure 47), and be directly linked to exposure. This is what is done for men exposed to radiation. Workers in industry where exposure to atomic or X-ray radiation is likely carry photographic plates that 'fog' at a rate directly proportional to exposure: they thus integrate the dose received at the surface of the target over whatever period is defined as the standard exposure time. For such radiation this is logical because effects can be exerted at the skin and the whole dose reaching the target can be regarded as 'absorbed'.

However, in many cases only a part of the pollutant to which the target is exposed is actually absorbed. Very often the relationship between exposure and uptake is not known: once absorbed, moreover, the distribution of a substance in the body can affect the issue a good deal. Some arbitrary judgements are inevitable, as when setting standards for food. It is believed, for example, that an adult man absorbs only about 10% of the lead in food and drink while children may absorb up to 50% (Chapter 3 and DOE, 1974b). Any primary protection standards for this metal would be expected to take into account the absorption level and allow twice as much of the metal in the food as would be tolerable on the 'inner' side of the gut wall. In contrast, but avoiding many problems about absorption, standards for lead in the industrially exposed worker in Britain and other countries are set inside the body rather than at the surface, allowing 80 μg/100 ml blood as 'acceptable' but removing men from contact with the metal above this point. Rather similarly, DDT and other persistent organochlorine levels in human tissue are monitored in most developed countries, as a watch on the exposure of the population,

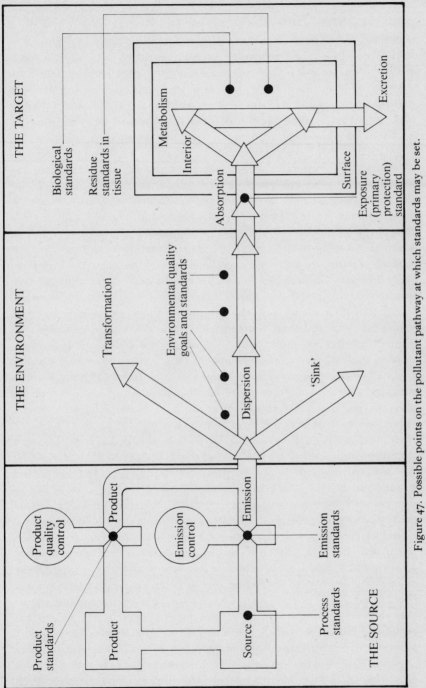

Figure 47. Possible points on the pollutant pathway at which standards may be set.

although there are no standards and it would be difficult to reduce exposures quickly. As Figure 46 (which illustrates an abnormally wide hypothetical variation in target response) stresses, however, there is a further question. The primary protection standard set needs to take account of the proportion of the population that must be protected from a particular level of effect. If, for example, no detectable effect at all is tolerable in any individual, only ten 'exposure units' would be permissible in the case illustrated, and a primary protection standard would accordingly be based on the worst case rather than the average exposure in the population. If, however, it is decided in the case displayed in the figure to tolerate biochemical changes but no physiological or behavioural impairment, a standard based on twenty-five 'exposure units' could be accepted (although again we would need to know the variation in exposure in the population). This is done in industry, where the standard for allowable lead in blood of workers is above the point (20–25 $\mu g/100$ ml) at which the activity of the enzyme delta-aminolaevulinic acid dehydratase in blood is impaired, and some neurological effects postulated (Chapter 5). If all that we were prepared to do was prevent death from acute toxicity in any element in the population illustrated in Figure 46, the standard would be set at about sixty-five exposure units – leaving over half the population with impaired competitive or reproductive performance and virtually all the remainder with physiological or biochemical changes. Finally, if all we were concerned about was to sustain the population we could allow that level of exposure which kept mortality within the limits that reproduction and recruitment could make good. This (although no standard was actually set) was in effect what happened for the heron population in Britain in the 1950s and 1960s.

In this process a common fallacy needs to be guarded against. It is that the information needed to set protective standards is gained by determining the *average* exposure circumstances, or the *average* level of target susceptibility. This is not so. Targets – especially living ones – vary in susceptibility because of genetic variation, variation in other factors predisposing to sensitivity or resistance to pollutants (a well-fed person may be more robust, for example, than a starveling), variation in age, variation in exposure to other pollutants, or variation in stress from quite separate causes. Before meaningful protection standards can be set, a programme of sampling is needed to define the variation in susceptibility in the population, and if possible to relate this to age and circumstances so as to explain why the most sensitive cohort has this property. A standard that genuinely protects 99.999%, or 99% or 90% of the population, or whatever proportion is decided on, can then be set – and if it is known why some elements are most sensitive special measures can be framed for their protection while allowing the more robust elements a higher exposure level.

This has been done in a few places in Britain in droughts, when the nitrate concentration in drinking water has risen temporarily to the point (11.3 mg N/l) at which bottle fed babies might be at risk from methaemoglobinaemia and special supplies have been delivered to the households concerned. Since protection costs money and social effort, it is rather important to assess needs realistically, in this way. It is also important to evaluate the level of protection that is likely to be attained.

It is probably sensible to use the term 'primary protection standard' for *surface* measurements of allowable exposure, or for measurements related to the surface. Examples are standards for radiation exposure, food and drinking water. The term *residue level standard* can be used for measurements internal to the body as with lead in blood or DDT in fat (although substances in blood are not, strictly, residues). A third standard could be envisaged within the target, based on its response: a *biological standard* which would (for example) define the maximal tolerable level of mutation in body cells due to radiation or the maximal tolerable impairment of delta-aminolaevulinic acid dehydratase activity in blood due to lead. (This is not the sense in which the term 'biological standard' is used in some publications, e.g. DOE (1977b). The latter defines it as 'levels of accumulation of a pollutant which must not be exceeded in a biological system'.) So far these biological standards have not been widely defined although theoretically they have much to commend them, at least as goals: they would not, however, be easy to link to meaningful legislation covering the general environmental field.

The 'exposure units' used in Figure 46 to illustrate how one can debate what the primary protection standard should be are helpful in forcing us to think about the simple fact that the target population is as variable in its response to pollutants as to any other external factor. But the arbitrary nature of 'exposure units' can also lead to another dangerous oversimplification. The effects of pollutants vary with the time of exposure and the precise pattern of that exposure. It is not wise to assume that the effect of a particular exposure level and period will be the same as that of twice the period and half the level. Nor is it valid to assume that the integral of exposure over a period is a sound measurement when all the evidence is that weighting must be given to peaks of exposure: that ten times the exposure for one-tenth of the time may be vastly more damaging than steady exposure to a unitary level. Much of the damage done to vegetation by photochemically produced oxidants is the result of short peaks. The pattern of variability in exposure needs to be established and related to the pattern of variability in target sensitivity.

It is fairly evident that a primary protection standard set to guard against acute toxicity must be based on short-term or *peak* exposure of a cohort of the population of known size and sensitivity. Food standards are commonly set well below such thresholds of acute toxicity, but above the levels that would be acceptable for continuous exposure, on the principle that because the standards specify *maximal* allowable levels, manufacturers, taking account of the normal variation in food composition, will ensure that a much lower *average* level is attained in their products. Since human purchase on the market involves random sampling and most people eat a mixture of foods, average diets will also contain levels of contamination well below the maximum allowable level. The process thus has a safety factor built in. The physiological processes of absorption and excretion, in which a continuous period of exposure is needed for body residues to build to a plateau level, will exert a further damping down effect. Biological and environmental circumstances here combine to smooth out exposure peaks. Standards set to prevent sub-lethal effects, in contrast, will generally be based on averages because such sub-lethal effects are commonly associated with chronic exposures.

Whatever the qualifications about the need to take account of variability in the target population, variability in exposure time and variability due to possible additive or synergistic effects, there is no doubt that primary protection standards are in principle the most logical of standards for they express just how much exposure, or dose, we consider it allowable for the target to receive. This does not mean that they are easily enforced. Indeed, the only means of enforcement are through modifying target exposure in direct response to the dose received (as revealed by monitoring) or through continuous screening of a pollutant source before it comes into contact with the target, for example, through checks on food quality. The latter process is really comparable to emission or environmental monitoring.

6.2.2 Environmental standards

At present, primary protection standards are not common. They are likely to be developed increasingly as understanding of the dose–effect relationships grows. The commonest kinds of standards are however set a 'step back' along the pathway from source to target, in the environment. These are sometimes referred to as *derived working limits* or *levels* (derived, that is, from a notion of allowable target exposure). One kind of derived working limit is an *environmental quality* or *ambient* standard, which is the maximum concentration of a pollutant in air or water allowed over a particular period.

Such standards are readily understandable because they define the wholesomeness of the medium to which a target is exposed. If they are based on soundly established criteria, they should guarantee that targets will not suffer unacceptable damage. For some time, figures of this kind have been set in industry, specifying the airborne concentrations of hazardous substances to which it is believed that working adults may be safely exposed through a 40-hour working week and a full working life. These *threshold limit values* are published by the American Conference of Governmental Industrial Hygienists and are adopted by the UK Factory Inspectorate and other national agencies (DOE, 1977b). It is more difficult to cater for the longer exposure periods and more variable target populations in the general environment. The concept has much to commend it: the problem is what levels to set, and how to devise means of monitoring and enforcement.

As Figure 47 and Chapter 3 emphasise, the environment provides many complex pathways for pollutants. From the point of emission they are dispersed and diluted: many substances interact with others present in air or water and are transformed into less (or sometimes more) troublesome materials: a proportion is removed by absorption or deposition into 'sinks' such as seabed or river bed sediments or into soil. If we consider an emission from a single point into a medium in which it is uniformly dispersed, the concentration at a point in the medium will vary inversely with the square of the distance from the source (though as Chapter 3 pointed out, canalisation in air or water currents partly reduces this dilution). If allowance for a constant loss by transformation and deposition is also made, the reduction in concentration with distance is accentuated. It is evident that the distance of target from source is all important in determining exposure – and that an environmental quality standard set with respect to a point at a given distance from a source will be exceeded on one side and undercut on the other! Moving the point of monitoring would thus adjust the picture of 'success' in meeting the standard.

In setting environmental quality standards, as in setting primary protection standards, the question therefore again arises of the degree of target protection sought. If *no* targets are to be permitted to exceed a certain exposure level, then the standard that is set must be determined so as to protect those individual targets that are nearest to the source, or in areas of greatest pollutant accumulation. It will be necessary to define the field of variation in exposure of the targets and set the standards to protect the most vulnerable examples. This is the philosophy behind a World Health Organisation study group's recommendations for an ambient air quality standard of 250 μg/m^3 of smoke particles and 500 μg/m^3 of sulphur dioxide (as 24-hour mean concentrations in air), these being the concentrations above which hospital admissions have been observed to increase in London (Royal College of Physicians, 1970; Lawther, 1973a, b). The recommendation is thus virtually for a primary protection standard, set to ensure that the worst cases in the human population are not endangered, but in a fashion applicable to the ambient air in the general environment. Comparison with Table 18 shows that more stringent standards are applied for sulphur dioxide in many countries.

The WHO Working Party that elaborated this standard went on to advise a long-term objective (not a standard) which is also interesting: an annual mean of 50 μg/m^3 of smoke and 80 μg/m^3 of sulphur dioxide. This, broadly, is the level at which lichen floras become impoverished (though the peaks within these means may be as important as the yearly levels). The adoption of these long-term objectives would thus do much to protect the most sensitive known targets – and as a consequence, be likely to safeguard coniferous trees, pasture grasses and crops. Again, what has been done is extrapolate from target sensitivity to the environment to which the target is exposed.

Similar quality standards have been set in various countries for carbon monoxide, oxides of nitrogen, ozone and other photochemically-produced oxidants, unburned hydrocarbons, lead in air, and a multiplicity of substances in water. Few of these are based on rigorous studies of dose– effect relationships at the target. All apply the same basic concept however: the determination of how much of a contaminant should be tolerated in the environment to which targets are exposed. As a concept it cannot be faulted. In Chapter 7 the difficulties it gives rise to in monitoring are examined.

It also poses problems in control. This is because there is simply no way of purifying the air as it blows over a city, or a great river as it winds through the plains. Once too much pollutant has got into the environment, it is too late to do more than rely on natural processes of dispersion and deposition in the 'sinks' discussed in Chapter 3. An environmental quality standard provides a goal to strive towards, a reference point against which to measure achievement or an indication of how much safety there is in hand. It is not as easy as it sounds to be sure that attainment in relation to it is being correctly measured and it is quite impractical to attempt direct enforcement.

The arguments rehearsed so far imply that an environmental quality standard depends upon an understanding of exposures and of target responses. The idea that 'our goal must be clean air' presupposes knowledge of how clean 'clean' is – and so far, 'clean' has been equated with 'adequately safe'. The alternative concept sometimes encountered, that *no* additional input of materials to the air, rivers, soil or sea over that which would be there in a non-industrial world is allowable, has a different philosophical base. It depends on the presumption that any deviation from the natural condition is an impairment of the habitat, and undesirable – or alternatively that we cannot predict the consequences of such a deviation and should be cautious. It is reinforced by the uncertainties in the process of risk assessment. Consequently (so this argument goes) we should deny ourselves the use of the capacity of the environment to assimilate wastes, refuse to allow any increase in contaminant concentrations, and wherever possible reduce existing levels. It was put particularly forcefully by Davis (1971) who said:

When damaging pollutants are identified, they must be stopped . . . No longer are we going to depend on nature's assimilative capacity to take care of the problem. Even in remote areas where discharges have been allowed, we will no longer allow it . . . We simply do not know enough about the science of ecology to predict the results, and we must be sure. The only way to be sure is to stop the pollutants at source – to keep them inside the factory fence.

The first approach incorporates the philosophy that 'the capacity of the environment to receive waste is a resource we can use but must not misuse' (DOE, 1972a). It assumes that the use of this capacity up to the point of unacceptable risk is a free good in economic terms, and that the purpose of the environmental quality standard or guideline (at least in part) is to say how large that 'good' is, In contrast, the more cautious approach explicitly sets out to avoid the use of that 'free resource'. If it is followed, the first step is to measure existing concentrations of substances in the environment, and where these are very low, to adopt them as environmental quality standards forthwith, irrespective of their relationship to target susceptibility. The next step is to establish programmes elsewhere which will either attain these 'natural' levels or move to an intermediate goal: in the latter case subsequent programmes would move progressively towards the background level. How far and how fast the process goes would still depend on social value judgements which need to balance the scientifically determined hazard of a contaminant level, the popular desire for the cleanest possible world, and the costs of attaining it, including the possible restrictions on industry, farming, or individual freedom. It is clear that the environmental quality standards inherent in this second approach are not statements about the resilience of environmental systems or targets.

Both approaches to environmental quality are manifest in most developed countries. In the United States the National Environmental Policy Act has been argued in the courts as demanding no deterioration in air quality when new industry is located in a rural zone. In the Paris Convention of 1974 a clause established as a goal that the pollution of the sea as a whole will not be increased by the relocation of industry (generally overall pollution is reduced by the substitution of efficient new plant for outmoded old equipment but relocation may raise levels of contamination around a new works). In Britain most Water Authorities have pursued a policy of never allowing the rivers in their care to deteriorate, and as they review consent conditions for discharges, insisting on progressive gains in quality. The Alkali and Clean Air Inspectorate are required by law to ensure that 'best practicable means' are used at all times to control pollution, and if new methods that will allow improved pollution abatement at acceptable cost come to hand they are generally adopted, regardless of whether the receiving air is already cleaner than an environmental quality standard set on rigorously measured target tolerances.

Wastes have to go somewhere. A prohibition on their disposal to the environment may be an incentive to recycling (which many people intuitively regard as desirable). But it is not a free incentive. If the disposal capacity of the world's air and waters is not to be used, even in a restrained fashion, the alternative may impose a heavy cost burden or mean that a particular industrial product is foregone. A compromise is commonly sought, using more than one set of figures to define environmental quality.

The first of these may be a standard or objective expressing the concentration that must not be exceeded in order to protect important targets. It is a basic health protection standard. The second figure indicates how far it is considered right to use the environment as a sink. It would define the resource available for waste disposal. The gap between them would be, from the point of view of targets, a safety margin. The two WHO recommendations concerning air pollution can be interpreted in this way. Ambient levels of smoke and sulphur dioxide might be brought below 250 and 500 $\mu g/m^3$ respectively as an urgent step to protect human health, but it might also be agreed not to allow clean air to be contaminated above the annual means of 50 and 80 $\mu g/m^3$ set in the long term objective, thereby using this as a definition of the extent of the legitimate use of the environment as a sink. This has been done in the United States, where a 'primary' air quality standard of 365 μg SO_2/m^3 (as a 24-hour mean) has been set to protect public health while secondary air standards have been established to safeguard vegetation (CEQ, 1973). In Britain, the Royal Commission on Environmental Pollution has suggested guidelines under which, between an upper level above which action to reduce pollutant concentration should be taken and a lower level below which action to reduce concentration would not generally be justified, 'target bands' of concentration might be set by control authorities. These bands would take account of local circumstances, allowing higher concentrations in an urban area (Royal Commission, 1976).

In a similar way, the lower of the two biological objectives for estuaries, proposed by the Royal Commission in its Third Report (1973), that the estuary should support a substantial number of living organisms in its sediment, might be considered a basic minimum quality standard or goal. It would not be a very high standard: muds substantially polluted with metals can still contain considerable invertebrate life. The Royal

Table 18. *Air quality standards for sulphur dioxide (averaged over 24 hours) as they were in 1973*

	Sulphur dioxide (μg/m^3)		Sulphur dioxide (μg/m^3)
Canada	300	Finland	250
Israel	250	Netherlands	275
Poland	350	Rumania	250
Sweden	250	Switzerland	
Turkey		(summer)	500
(industrial areas)	300	(winter)	750
(residential areas)	150	United States	365

From CEQ (1973).
Note: In addition to requiring that these figures should not be exceeded as an average over 24 hours, all these countries also stipulated that the concentrations should remain below the standard for 95% of the time.

Commission's other guideline – that migratory fish like salmon should be able to pass up an estuary and river at all stages of the tide – might be taken as a goal, to be sustained where this degree of quality already existed, agreed as a measure of the maximum use of the system as a sink, and pursued as opportunity allowed where conditions at present fall short of it. In each case a third figure, expressing what was supposed to be the condition in a pristine world, might be calculated as an ultimate goal or reference point.

This approach is discernible in recently issued European Communities' documents on water standards for fresh-water fish, although as a House of Lords' Select Committee on the European Communities points out, the details are somewhat confused (House of Lords, 1977). It is proposed to define rivers as *salmonid* (i.e. capable of supporting salmon, trout and comparably demanding fish species) or *cyprinid* (i.e. capable of supporting coarse fish like roach or perch). The water qualities specified are however more stringent than is necessary to achieve these biological goals, possibly because of the intention to allow a large safety margin.

A characteristic of this system of standards or objectives is that the first figure in each of these groups is likely to be fairly non-controversial, and readily justified because it relates to acute effects on human health or other important targets. Table 18 sets out such standards for sulphur dioxide as they were in various countries in 1973. The more stringent standards or objectives are, on the other hand, more susceptible to argument. There may be doubts about the value put on sensitive plants or ecosystems. Equally, if some rivers are already clean enough to support healthy coarse fisheries, it may be considered better to spend resources on cleaning others which are at present grossly polluted rather than make further

Table 19. *Water quality values proposed in 1963 by the heads of the Water Management Authorities of Member States of the Council for Mutual Economic Aid*

	Unit	Class I	Class II	Class III
Dissolved oxygen	(O_2) mg/l	> 6	> 5	> 3
		The results of night and morning samples not to be taken into account		
Dissolved oxygen (% of saturation)	(O_2) %	> 75	> 50	> 30
		The results of night and morning samples not to be taken into account		
BOD_5	(O_2) mg/l	> 5	> 10	> 15
Oxidability permanganate value, $KMnO_4$	(O_2) mg/l	< 10	< 15	< 25
		Excluding waters containing humic substances		
Free hydrogen sulphide	H_2S mg/l	ND	ND	< 0.1
Biological condition, saprobity	oligo-beta-beta-meso alpha-meso			alpha-meso
Chloride ions	Cl^- mg/l	< 200	< 300	< 400
Sulphate ions	SO_4^{2-} mg/l	< 150	< 250	< 300
General hardness	German degrees	< 20	< 30	< 40
Calcium ions	Ca^{2+} mg/l	< 150	< 200	< 300
Magnesium ions	Mg^{2+} mg/l	< 50	< 100	< 200
Dry residue of matter in solution	mg/l	< 500	< 800	< 1200
Suspended matter in flow in dry weather	mg/l	< 20	< 30	< 50
Ammonium ions	NH_4^+ mg/l	< 1	< 3	< 10
Nitrate ions	NO_3^- mg/l	< 13	< 30	
pH	—	6.5–8.5	6.0–8.5	5.5–9.0
		Excluding naturally acid waters		
Total iron	Fe mg/l	< 0.5	< 1	< 1.5
		Excluding waters containing humic substances		
Manganese	Mn mg/l	< 0.1	< 0.3	< 0.8
Volatile phenols with water vapour	mg/l	< 0.002+	< 0.02	
		+ in case of chlorination of drinking water		
Detergents (active washing substances)	mg/l	< 1	< 2	< 3
		Only for anionic washing substances. For other substances special maximum limits must be established where appropriate methods of analysis are available		
Cyanide ions	CN^- mg/l	< 0.01	< 0.02	< 0.1
		Take into account compound cyanides as necessary		
Temperature	°C	Each country should establish its own depending on climatic conditions		
Smell and taste	—	Not noticeable	Not out of the ordinary	At the most only slightly out of the ordinary

	Unit	Class		
		I	II	III
Colouring	—	At the present time indication of exact data is impossible on account of analytical difficulties		
		Can be included in the classification depending on given circumstances		
Oils	—	Invisible	Traces only	Traces only
E. coli titre	—	0.1	0.01	—
		In determining the general *coli* content in accordance with the lowest titre values		
Pathogenic microbes	—	ND	ND	ND

From WHO (1966).
Notes: (1) 'ND' means 'not detectable', (2) class I is water for public supply, the food industry, trout and salmon fisheries and bathing; class II is water for fisheries other than trout and salmon, sport and amenities, and livestock watering; Class III is water for irrigation and industrial use.

improvements to bring the first group up to salmonid standard, or better (House of Lords, 1977). The value of the waste disposal capacity of the environment is equally likely to vary within and between countries: if a nation is small, densely populated and in urgent need of development, the balance of advantage may lie in allowing the quality of parts of the environment to shift towards the primary or 'basic health' standard. How far this shift is tolerated depends on judgements of the value of environmental waste disposal, wildlife and amenity, short of the limit of manifest hazard to human health, livestock and crops and other vital resources.

The standards in Table 18 are legal stipulations, more or less firmly enforced. Table 19, in contrast, sets out guidelines indicating the quality of a medium (in this case water) considered desirable for various uses when they were drawn up, in 1963. Later international and national measures include other figures (House of Lords, 1977 and Tables 20, 21) but the broad principle of defining the quality of water appropriate for different uses remains.

6.2.3 Emission standards

Whatever environmental quality standards are set, they cannot be attained except by control of emissions to the environment. The third kind of standard is that set for such discharges – *emission standards*. Their aim is the same as for the other kinds already discussed – to protect targets – but they are two steps back from the target. They are the means whereby

Table 20. *Water standards in Canada and the United States relating to the Great Lakes*

Parameter	Waters	Standard
Dissolved oxygen	Upper waters	Above 6 mg/l at all times
	Lower waters	Not less than necessary to support cold-water fish
Coliform bacteria	General	Below 10000/l
	Local	Below 2000 faecal coliforms/l
Dissolved solids	Lower lakes	Below 200 mg/l
Iron	General	Below 0.3 mg/l
Phosphorus	General	Levels below those permitting nuisance growth of algae, etc.
Temperature	General	No change that would adversely affect local or general use of the water

From Davis (1971).
Note: the difficulty of establishing numerical values for target response is evident from the 'standards' listed here for phosphorus and temperature. These 'standards' are in fact quality objectives towards which the authorities concerned agreed to work.

environmental quality standards are achieved and primary protection standards therefore secured.

It is fairly obvious that if the aim is to achieve a certain level of environmental quality the same emission standard cannot be applied to every discharge. In an area where the level of a pollutant in air or water is already at or even above the maximal allowable level, no new discharges at all are tolerable unless existing emissions are reduced. In an area where the levels are well below the environmental quality standard, relatively large emissions may be permitted – assuming one is adopting the philosophy that the dilution capacity of the environment is a usable resource. Clearly if the aim is to sustain a certain level of environmental quality, therefore, different standards must be set for different emissions, the control applied to each being calculated in relation to the capacity, or degree of saturation, of the environment at the point of discharge. Generally speaking, controls will be most stringent in areas already near the acceptable limits of pollution, where there is a high density of emissions, or on large emissions liable to give rise to very high concentrations in their vicinity. Laxer standards can be applied to widely-spaced, small emissions in areas of clean environment. Figure 48 illustrates the principles involved.

Proposed new sources of emission A–E are sited in areas of clean environment, and could in consequence be allowed to emit relatively large amounts of material (over 4 arbitrary 'units') whereas F–K are in an area

Table 21. *Proposed European Communities standards and guidelines for waters supporting fresh-water fisheries*

	Standards or guidelines			
	Salmonid		Cyprinid	
Parameters	Guideline	Indicative	Guideline	Indicative
Temperature (°C)	—	Summer ≤ 20 Winter ≤ 10	—	Summer ≤ 25? Winter ≤ 10
Dissolved oxygen (mg/l)	50% 9	50% 9 95% 7	50% 8	50% 7 95% 5
	100% 7	100% 5	100% 5	100% 3
Suspended solids (mg/l)	25	—	25	—
pH	—	6–9	—	6–9
BOD (mg/l)	3	—	6	—
Phosphate (mg PO₄/l)	0.2	—	0.4	—
Nitrate (mg/NO₃/l)	3	—	6	—
Total ammonium (mg NH₄/l)	0.04	1	0.2	1
Nitrite (mg NO₂/l)	0.05	—	0.5	—
Monohydric phenols (mg/l)	—	0.005	—	0.005
Non-ionised ammonium (mg NH₃/l)	—	0.005	—	0.005
Zinc (mg/l)	10	—	—	—
	50	—	—	—
	100	—	—	—
	500	—	—	—
Water hardness (mg CaCO₃/l)	—	0.03	—	0.3
	—	0.2	—	0.7
	—	0.3	—	1.2
	—	0.5	—	2.0

Note: (1) The proposals headed Indicative would be binding: those headed Guideline would be open to variation by member states. (2) The draft directive also specifies the methods to be used in measuring the parameters listed, and the frequency of sampling. (3) The House of Lords Select Committee on the European Communities (Environmental Policy) pointed out that a number of these standards appear unduly restrictive since fish flourish in waters that fail to meet them (House of Lords, 1977).

with diminishing capacity and a need for tighter standards, and L, M and N are in a saturated zone where no further emissions are tolerable unless the safety margin is abandoned. New discharges would be possible only if existing sources were improved, thereby creating capacity for new industry.

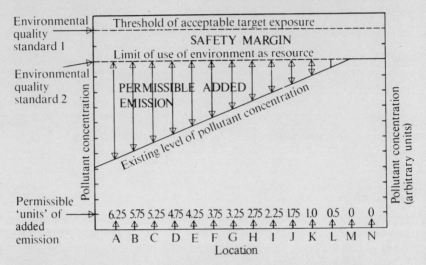

Figure 48. Diagrammatic relationship between environmental quality, existing pollution level, and unused environmental dispersion capacity.

It is sometimes argued that uniform emission standards should be adopted throughout a country or internationally. Generally three reasons are advanced for this: environmental, practical and economic. The environmental argument is based on the premise that the tightest possible controls should be imposed everywhere in order to keep the environment as clean as possible. This is a perfectly tenable viewpoint, so long as it has been agreed to reject the environment's pollutant disposal capacity as a resource, and work toward a baseline of minimal contamination. It would evidently be necessary, on this principle, to adopt everywhere the most stringent possible emission control. Using Figure 48 as an example, and assuming that the best practicable means for emission control still led to one 'unit' of contamination, this would still imply that plants L, M and N should not operate, while all the others would have to operate to the standard demanded of emission K. The result would be 11 'units' of emission instead of the 41.75 'units' tolerated if differential standards were applied and the environmental capacity fully used. This universal adoption of the most stringent standard is fairly clearly the most costly solution. The universal adoption of a less stringent standard would equally clearly be unsatisfactory if it allowed the regions around plants F–N to deteriorate below the basic health standard.

The practical argument for adopting uniform emission standards is that this is simplest. It avoids the difficult computation of how much emission an area of environment can safely take. This calculation is complicated because of the need to consider climatic variation – an emission from a factory to air, for example, must be much better controlled when there is a pool of stagnant air trapped under a temperature inversion about the point of release than when the plume of pollutant is passing straight into a full gale. For rivers, likewise, emission control linked to dilution capacity needs to take flow rate into effect, or environmental damage will occur at times when the river is low. The emission standards need therefore not only to be different for every discharge but also to vary from time to time according to conditions, and this demands very sophisticated controls. It may appear easiest to set a uniform, rigorous standard to be met at all times from all examples of a particular kind of factory or piece of equipment.

The third argument is economic. It is often said that to impose different standards on factories making the same product is to distort trade: to give competitive advantage to one rather than another and to impose a degree of 'unfairness'. This argument is especially heard internationally. It is supported by the statement that if less stringent standards are imposed on a factory in Bergen, Vancouver or western Scotland than on one in the Ruhr or Los Angeles, the production costs of the former are lowered. Opponents of this argument point out that there are many variables that impose cost differentials upon industry – distance from market, distance from raw material sources, availability of labour, availability of public transport, distance from ports, mean annual temperature (affecting heating costs), energy costs and the like. They argue that it is illogical to try to make the pollution control cost uniform when all these others vary. It is also wasteful in terms of resources, if this means that a factory is prevented from using a legitimate natural resource – environmental waste disposal capacity – which may in a remote site offset some of the higher costs of isolation. Faced with these uncertainties and grounds for debate, it is not surprising that there is a considerable variation in how emission standards are actually fixed, even within a single country. In Britain Water Authorities impose consent conditions on individual discharges to rivers (and when the 1974 Prevention of Pollution Act is fully operative they will also do so on discharges to estuaries and coastal seas). These consent conditions are fixed according to the nature of the individual emission and the condition of the receiving waters (in other words, emission standards, even applied to the same industrial process are not the same throughout the country or even river basin). The current condition of many of the British rivers into which industrial effluents are discharged certainly makes this tight control, and its progressive improvement as consent conditions are reviewed and new methods of pollution control become available, a defensible approach to a goal of enhanced environmental quality. In the United States permits have been required since 1971 for all industrial discharges to navigable rivers and these too have been operated to meet water quality standards. Despite progressive improvements, British rivers, as a resource still appear generally over-used (DOE, 1970, 1972) although the majority of river reaches are below the threshold of unacceptable damage, lying in the 'safety margin' of Figure 48. The same generalisation probably applies in the industrial zones of Western Europe and North America.

Discharges to air in Britain from the main polluting industries are fixed by the Alkali and Clean Air Act Inspectorate on the basis that 'best practicable means' shall be used to control emissions from every plant. The exact nature of 'best practicable means' varies from process to process. It takes account of what is at present technically feasible and economically acceptable. It tends to result in uniformity of emission standards over the country for a particular kind of plant or process (a number of examples are given in HSE, 1975). In addition, the nature of the environment at the point of emission is considered, and chimney height altered to ensure adequate dilution in each case: a blend of basically uniform emission standards and variable dispersion to ensure environmental quality.

Mobile sources of emission such as motor vehicles must, it can be argued, have uniform emission standards for they could otherwise go from areas in which the air is clean and vehicles few and hence 'dirty emissions' tolerable, to congested cities where tight standards are essential. This is the case for European Community standards, based on those developed by the Economic Commission for Europe. Such standards are inevitably 'town standards' – that is, they are based on the need to sustain environmental quality in the 'worst case' of an urban zone. Rather similarly the United States' vehicle emission standards were based upon the serious problems in Los Angeles and other large cities. There are vehicles in North America (for example in rural Canada) – or in the outskirts of Europe (such as the Western Isles) which will cost more because of urban pollution control devices they do not need, and hence waste resources. It is generally argued, however, that flexibility of standards for such sources would be impracticable.

6.2.4 *Other types of standard*

Emission standards are sometimes attained by direct filtration of an emission, for example through an electrostatic precipitator, a bag filter, a wet washing system or a sewage works. At other times they are attained by controls on a manufacturing process so that pollutants are not produced in unacceptable quantities. Such controls on processes can be termed *process standards*. Ideally they are fully compatible with production efficiency. Well-maintained and well-regulated diesel engines run economically and do not emit black smoke. Water economy is easier if a process does not emit an effluent demanding great dilution. Many industrial emissions are controlled by the Alkali Inspectorate on the basis of process controls. They act at the most immediate point on the pathway from source to target at which meaningful standards can be set. If they are wisely judged, they may be most economical of all because they prevent the creation of pollution and make subsequent emission treatment or target protection unnecessary.

Process standards are in many ways analogous to *codes of practice*, widely applied to such activities as the spraying of pesticides, the application of fertilisers (especially from the air) or the disposal of solid waste on land or by dumping at sea. Such codes of practice govern what is put where in what amounts and under what environmental conditions. They are closely linked to environmental condition and disposal capacity: the whole process of waste disposal on land or at sea assumes as a premise that a hole in the ground or an ocean deep can legitimately be used as a resource for this purpose.

In a sense, the decision about the desirable level of protection from radiation may be regarded as leading to such a code of practice. The International Commission of Radiological Protection in 1973 (ICRP, 1973) promulgated a recommendation that:

As any exposure may involve some degree of risk, the Commission recommends that all unnecessary exposure be avoided and that all doses from justifiable exposures be kept as low as is readily achievable, economic and social considerations being taken into account.

The assumption is that maximal attainable protection is needed: that no standard as such can be stated quantitatively, and that the emphasis should be placed on steps to minimise exposure – a code of practice. The requirement in UK industrial air pollution prevention that 'best practicable means' be employed to reduce polluting discharges is another 'code of practice' in this sense, leading to a variety of emission controls.

Finally, *product standards* are a variant used to regulate the degree to which a manufactured article is liable to cause an environmental hazard as a side-effect. These standards are very numerous. Motor vehicles are designed to meet standards of exhaust emission. Ship's sanitary systems are increasingly being designed so as not to discharge raw sewage into harbours or inshore waters. Tankers are being designed to prevent unacceptable spillages of oil – and often have segregated ballast systems so that water taken aboard to trim a vessel when travelling unladen does not become contaminated, as it does when ballast water is carried in empty oil tanks, causing pollution if it is pumped out again. Pesticides, pharmaceuticals and food additives are developed to meet criteria of effectiveness in use, but are being screened by increasingly stringent tests to limit unwelcome side-effects. Detergents in Britain are now required to have a certain minimal biodegradability so as to prevent foaming of rivers when the waste waters that contain them are discharged, and it is proposed to extend these standards throughout the EEC.

6.3 The inevitability of change and variation in standards

Standards are codified statements about allowable performance or allowable exposure. All relate ultimately to the degree of risk accepted for a target, and all therefore include an element of value judgement. The degree of protection demanded varies with the nature of the target and of the human community concerned. Primary protection standards for man differ from those that would be adopted for farm livestock or for wildlife. The primary protection standard is the first link in the chain. Environmental quality standards are expressions of exposure levels tolerable in the environment as a safeguard against exceeding a primary protection level. They may go further and incorporate value judgement about goals of environmental purity only slightly related to target protection. They may enshrine both concepts of essential protection for targets and judgements about the extent of the legitimate use of the environment for waste disposal. Emission standards define what may be released to the environment if environmental quality limits are not to be exceeded. They in turn demand process standards for industry, product standards for many articles, and codes of practice in many operations. The whole system is an interlocking series of statements about social demands for environmental protection. By themselves the standards regulate nothing: it is the practices adopted to meet them and the regulations that ensure that those practices are enforced that really count. Environmental health can be ensured without standards, if the practices are right. But standards are often useful direct expressions of what a community is trying to do.

From all the preceding analysis it is obvious that no particular series of standards should be looked on as final. Research continually reveals more about the susceptibilities of targets, affecting the primary protection levels required. Technology continually generates both better instruments for monitoring and better means of abating pollution at economic cost. Shifts in the wealth of nations and the values people put on the environment demand adjustments in the quality they seek and can afford. Because standards are statements about environmental policy, they must be expected to be adjusted, within the total context of national and international knowledge, capacity and desires.

Finally, it is important to discriminate between standards as statements of desirable dose or exposure limits – that is, standards as expressions of scientific logic tempered by economic and social judgements about the benefits of polluting activities and the social costs of the resulting harm – and standards as legislative instruments. Once a standard has been determined, how it is enshrined in legislation or administered by an appropriate control agency is a matter for international, national or local choice. So long as it is agreed and pursued effectively a standard may not, indeed, need to be backed by legislation at all. It may be quite adequate to state it as a social goal or objective and rely on competent adminstrative agencies to pursue it using the best means at their disposal. Standards are not needed in order to create sound legal and admistrative procedures for pollution abatement: nor is the analysis of dose and likely risk and the establishment of desirable objectives for target protection pointless even if it is not immediately backed by law or international convention. It helps to have analysed the situation and gained a mark to aim at.

7. Monitoring and surveillance

7.1 The purpose of monitoring

If an effective system of environmental management is to be designed, information is needed about:

(a) the substances entering the environment, and their quantities, sources and distribution;
(b) the effects of these substances;
(c) trends in concentration and effect, and the causes of these changes;
(d) how far these inputs, concentrations, effects and trends can be modified, and by what means, at what cost.

The first kind of information is provided by survey, defined as a programme of measurement that defines a pattern of variation of a parameter in space. Such surveys indicate the scale of pollution at a particular time. But to interpret these patterns and assess risks, research on the effects of the pollutants is needed, leading where appropriate to standards and regulatory measures. Such regulatory measures are continuing social acts related to the continuing discharge of wastes to the environment; no static survey can therefore meet our need for 'feedback' to a management system. What is needed is a system of measurement of change over time: monitoring and surveillance.

The place of monitoring within the chain of actions leading from the initial recognition of a problem to the imposition of controls has been stressed by SCOPE (1977), whose report is largely based on the work of the Monitoring and Assessment Research Centre (MARC) at Chelsea. Figure 49 is adapted from that report, but differs in several important respects from SCOPE's approach. Successive steps of identification of a problem or need, analysis of information about it and response in terms of further data collection or action to deal with the problem are inevitable. However, in Figure 49 a distinction is drawn between the process of *survey, surveillance and research*, designed to find out more about a problem before a policy is decided on, and monitoring initiated after the policy is established to find out how well it is working.

This accords with the definition of surveillance as the repeated measurement of a variable in order that a trend may be detected. By the

164

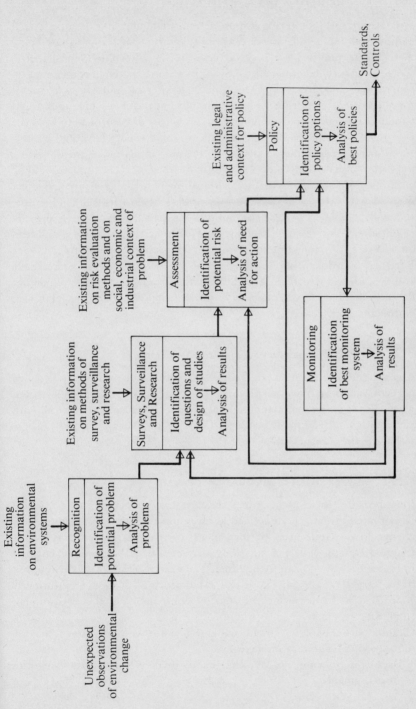

Figure 49. Place of monitoring in the context of problem recognition and social response.

same definition, monitoring is more action-orientated: it is designed to tell us how well our regulations are working and how far standards are being met. The frontier is hazy, and the terms have not been used consistently. Very often surveys designed to give a picture of the environment at one time are deliberately repeated after an interval and provide useful surveillance of trends. Equally commonly, surveillance defines a trend which is viewed seriously enough for measures of control to follow: the continued surveillance becomes monitoring and is supplemented by new monitoring programmes at points of emission or where targets are known to be most at risk. The whole complex of measurements is intimately linked with research, seeking to define relationships between exposure and effect more exactly. Moreover, as Figure 49 stresses, the model is not of a simple sequence. Policy can lead to monitoring which can in turn provide essential data for research seeking to explain effects and causes, and lead on to a re-evaluation of policies, with re-design of the monitoring system. For these reasons, while the theoretical distinction between surveillance and monitoring is useful, it is difficult to sustain, and in the remainder of this chapter 'monitoring' will be used in the more usual sense, embracing both concepts.

7.2 The design of a monitoring scheme

Any programme of survey, surveillance or monitoring inevitably demands decisions about: (*a*) what to measure, (*b*) where to measure it (that is, with what spatial pattern of sampling), (*c*) when to measure it (that is, with what pattern of sampling in time), (*d*) what method to use and (*e*) how to handle the information.

7.2.1 What to measure

Very broadly, these programmes measure two kinds of thing: (*a*) actual or potential targets (here defined as anything that may be liable to show change in distribution or performance), (*b*) factors, or potential factors (defined equally widely as anything that may be liable to cause changes in the environment or in living targets). These two classes will be referred to as 'target monitoring' and 'factor monitoring'. Target monitoring is concerned with measuring the state of components of the environment (including, for convenience, the media whose changes were reviewed in Chapter 4), or of living organisms and structures. Many targets may be monitored, as Table 22 indicates.

Factor monitoring, on the other hand, involves three main categories of measurement:

Table 22. *Target monitoring*

Target	Measurement
Physical systems	
1. Atmospheric systems	Circulation, composition
2. Estuarine and marine systems	Circulation, composition
3. Fresh-water systems	Circulation, composition
4. Land systems	Pattern, composition
5. Structures and materials	Corrosion, wear, soiling
Biological systems	
1. Ecosystems	Distribution, composition, overall performance (major fluxes)
2. Species and populations	
(a) Man	Distribution, performance (includes epidemiological data)
(b) Crops (including garden and forest crops) and livestock	Performance and distribution
(c) Wild life	Performance and distribution
(d) 'Indicator species'	Performance and distribution
3. Individuals of target or indicator species	Physiological and biochemical parameters, indicators of performance, residues as aids in analysis

(*a*) of physical attributes of the environment liable to affect targets (e.g. temperature, radiation, turbidity, cloud cover etc.);

(*b*) of chemical variables with similar potential effect (e.g. sulphur dioxide, particulates, hydrocarbons or oxidants in air and metals or hydrocarbons in water);

(*c*) of biological variables liable to affect other organisms or the physical environment (e.g. disease organisms, herbivores, predators, decomposer activity).

Too sharp a boundary should not be sought between target and factor monitoring. Changes in the performance of a target, especially if it be a physical target like a component of the atmosphere or ocean, can become a factor of major significance to a wide range of secondary targets (as explained in Chapters 4 and 5).

As with primary protection standards and environmental quality standards, the basic dichotomy between target and factor monitoring arises because of the nature of the pathway linking the release of a pollutant or alteration in a physico-chemical environmental variable with a target which may be distant in space or time (Figure 50). Measurement of chemical and physical factors in the environment, along the pathway, can usually be

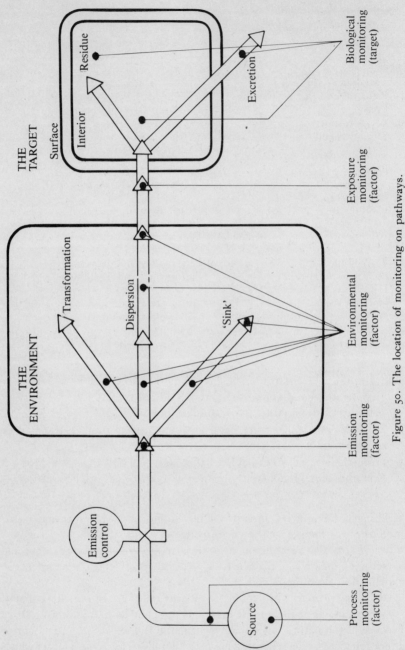

Figure 50. The location of monitoring on pathways.

related fairly directly to sources. Thus the monitoring of sulphur dioxide in air allows ambient levels to be related roughly to the distribution of power stations, domestic chimneys or heavy industry. If such measurement defines the pattern of variation in the concentration or intensity of a factor over the country, and if exposure–effect relationships are also known, it is possible to predict the likely pattern of effects on particular targets at particular places, identify areas of worst risk, and set standards in a way already described.

Conversely, monitoring of the performance of targets focuses on the things we are trying to protect and may tell us fairly immediately whether we are succeeding. The trouble with target monitoring is that biological targets respond continually to the interplay of so many variables, and have a homeostatic capacity to damp down some of their impact: a change in the distribution and performance of ecological systems or organisms cannot always be linked to a particular physical or chemical cause. Target monitoring may detect change and provide a warning that something is amiss: research may be needed before we are able to identify the cause and take protective measures. The consequence is that we commonly employ both types of surveillance and monitoring, but that the design of the monitoring systems differs fairly markedly.

Such activities must all have a purpose: in the present context, the provision of essential 'feedback' to environmental management. It is inevitable that monitoring schemes embrace a wide range of both factors and targets. Nonetheless, many more things could be measured than can ever be included in a practical scheme: selection is of vital importance. Practical schemes are likely to be concerned with the following categories:

1. *Targets* there are national or international policies to protect – (*a*) man, (*b*) livestock, crops, forests and fisheries, (*c*) wildlife, (*d*) landscapes of high 'amenity value', (*e*) structures or materials.
2. *Targets* which could serve as useful 'indicators' of threats to category 1 targets – (*a*) 'sentinel' organisms showing changes in advance of changes in priority targets, (*b*) 'accumulator' organisms with high residues in their tissues of toxic substances which could endanger priority targets or systems.
3. *Factors* known to affect priority targets, and known to be emitted or modified by current human activities.
4. *Natural environmental factors* known also to affect targets, and needing to be kept under scrutiny as part of the background without which the interpretation of monitoring becomes impossible.
5. Substances known to be emitted in quantity to the environment by man, as yet without proven effect but demanding precautionary surveillance as *potential factors*.

Right at the outset, attention must be given to the *methods* available for monitoring and the resources required. It is much easier to think of a parameter to examine than to establish the capability to do the work and to carry it out. Not only is there a need for valid design of environmental sampling systems, but even when this has been done, it does not follow that the results will be meaningful unless inter-comparable techniques of measurement are used throughout the scheme, and the reliability of the results is properly established. An examination of this problem by the Air Management Research Group of OECD (1972) drew attention to concern over:

(a) general measurement capabilities for very low concentrations of nearly all pollutants,
(b) comparability of measurements and reference methods,
(c) reliability and calibration of automatic instruments,
(d) sampling methods for stack gases and dusts.

The first three are general considerations equally applicable to aquatic pollutants or to residues in organisms.

Methods now exist for detecting many substances at very low concentrations (sometimes it is advances in such methodology, detecting substances at greater dilutions than was previously possible, which has given the appearance of expanding pollution: this may be true of the appearance of organochlorine pesticides in the Antarctic in the 1960s following advances in gas–liquid chromatography). The intercomparability of such methods, however, even between established laboratories, is not always as good as it should be. Holden (1971) reports the results of an OECD study group which exchanged samples of birds and marine organisms among laboratories in many countries: some initial large discrepancies were revealed, but reliability steadily increased and inter-laboratory and inter-method differences became far less than the variability of the material, justifying an international monitoring scheme. Such a pilot study, establishing that sampling, measurement and recording techniques of adequate quality exists, is clearly sensible before major programmes are begun. The adoption of reliable and intercomparable methods will be helped by a Working Group of SCOPE (Gallay, 1976) which has published a manual of both simple and sophisticated techniques for measurement of many common pollutants.

7.2.2 *Where and when to make measurements*

Factor monitoring can be done at many points along the pathway from the source of a substance to the target, as Figure 50 suggests. In particular it can be:

(a) within an industrial process (*process monitoring*),
(b) at points of emission to environment (*emission monitoring*),
(c) at a network of points in the environment (*environmental monitoring*),
(d) at the surface of a target (*exposure monitoring*),
(e) within a target (*internal monitoring*).

It is evident that if standards or controls have been imposed on the discharges from a plant, monitoring of individual emissions is essential to determine whether they are being observed. Once the emissions are mingled in the environment it may be quite impossible to pinpoint the source of excessive ambient levels. Monitoring *emissions* is unambiguous in terms of what is being sampled. This is why the Alkali and Clean Air Inspectorate and Water Authorities in Britain and comparable agencies in other countries, ask for it on discharges or emissions under their control.

For a manufacturer, it is small comfort to know that an emission is above the set standards after it has passed through the outlet to the environment. If the manufacturing process is to be regulated, *process monitoring* on the pathways to the points of emission may also be needed. Most modern plant has sophisticated feed-back systems to a central control which reacts to any malfunction of any component including pollution arrestment devices such as electrostatic precipitators or bag filters.

Process and emission monitoring thus indicate what is getting out. But standards are imposed because of what that emission will do, and this involves concepts of its behaviour in the open environment and on its way to the target. *Exposure monitoring* at the surface of the target is the other unambiguous point on the pathway which can reveal whether emission controls are or are not leading to satisfactory protection. With some substances and targets exposure monitoring is straightforward (e.g. the photographic plate system to detect exposures of workers liable to exposure to radiation). *Internal* factor monitoring may be needed to measure exposure, as with the routine measurement of lead levels in the blood of workers in this industry in Britain by the medical officers of the Department of Employment. A third example is the routine monitoring of contaminant levels in food, done in Britain by the Laboratory of the Government Chemist and the Ministry of Agriculture, Fisheries and Food (MAFF, 1971, 1973).

Target monitoring, in contrast, can only be done at or within the target, defining its overall response or the response of some component. It differs, also, from factor monitoring in that while the latter begins with a presupposition that the factor is important, and is conducted so that predictions, based on criteria, can be made about the likely hazard or the efficacy of controls checked, target monitoring begins at the 'other end'. It records changes in a target without initial or necessary certainty as to the cause. In all cases, however, the identification of cause should be the immediate next step once changes in a target have been detected, and this leads on to the establishment of criteria and the calibration of target responses against environmental change.

7.2.3 *Some characteristics of factor monitoring schemes*

Theoretically, process monitoring, emission monitoring and some exposure monitoring can involve 100% sampling – that is they can record continuously every source of emission or the exposure of every target. This is done for the most hazardous processes and emissions – like those from nuclear treatment works – or for the most dangerous human exposures like those to radiation or to certain toxic substances at work. These measurements, being individual, record a degree of variation in exposure as well as allowing average values to be derived. However, they do not necessarily give a full record of variation in time. Blood lead levels reflect the balance of intake and excretion over a considerable period: radiation monitoring likewise records a cumulative dose. They do, however, allow the determination of the 'worst case' target, and of how many potential targets fall into various categories of hazard.

The other type of exposure monitoring is not individualised. It is designed to give a valid indication of the *average* exposure of the population. It is exemplified by how food contamination has been monitored in Britain. Samples of diet have been obtained, from randomly chosen shops scattered over the country, the proportion of contaminant in each determined, and the results weighted according to the proportion that foodstuff makes up of average diet. The results of such surveys are valid as an indication of the exposure of the hypothetical 'average man', but say little about extreme exposures of the faddy who eat only one or two foodstuffs, or the unlucky who get them all from just one or two areas that are above average in contamination. The sampling system could, however, be extended to determine the *variation* in contamination levels in a type of food and, using a parallel system of monitoring dietary habits and their variance, to work out the probable extreme 'worst cases' in minorities. This has only been done in a few cases (including the determination of maximal exposures to the radioisotope ruthenium-106 among eaters of 'laver bread' made from the seaweed *Porphyra*). New studies to ascertain such ranges of exposure to other pollutants are proposed and a recent study of mercury in fish was designed so that both average and extreme exposures could be computed (MAFF, 1971).

In between the relatively unambiguous monitoring of two ends of the pathway, lies the environment. Here the design of monitoring systems is much more difficult.

The environment receives large numbers of emissions: not all the sources are recorded. The air or water flow mixes and dilutes them. Some interact chemically to produce further substances. Some settle onto soil, are trapped by the foliage of trees, are washed down in rain or are filtered by water plants and bound chemically or physically to the mud on the beds of rivers, lakes or sea. To obtain a meaningful picture of what is happening demands very careful sampling. The system must be designed to do two things. It should indicate where and when target exposure is likely to be above the desired levels (standard or goal) – hence it must define variation – and it should be associable in some way with emission monitoring so as to provide a measure of the adequacy of emission controls. If the smoke and sulphur dioxide levels in London are found to exceed the World Health Organisation limits on thirty days in the year, it is helpful to be able to ascertain the relative contribution of domestic or industrial sources, or of a factory complex in a particular area, as well as the significance of weather changes. Finally, the study should help to construct, test and extend models predicting the location and scale of likely pollution incidents.

The first need in sampling levels of an actual or potential factor in the environment is therefore to define the range of variation it displays both in space and time – for it has been pointed out in Chapter 5 that peaks of exposure to high levels may be more damaging than long exposure to lower levels of a toxic substance. How is this pattern of variation to be detected? First, it must involve a spatial pattern of sampling which is objectively determined. It certainly must not be done by locating the sample points according to some intuitive guesswork about where conditions are likely to be worst – or it may be impossible to decide afterwards how far the results have been biased by the procedure of 'trying to get the answer you first thought of'. If we have a hunch that most of the cats in Cumbria are black, it is no good trying to test the idea by deliberately concentrating the sampling in those areas where we have reason to suppose black cats are especially cherished by maiden aunts.

There are two approaches that could be adopted. One would be simply to place sampling points at random (using a computer to generate the national grid co-ordinates of points on the surface of a country). The other would be to first stratify the country into sub-zones and then to sample these as subsidiary components of the survey. One stratification might be related to urbanisation: it could be arrived at by recording for every 10-km grid square on the map the proportion of land shown as 'built up' and the number of roads and railway lines cutting the margins of the square. The computer could then be used to arrange all the squares in order of urbanisation and density of communications networks, and then to chop this series into eight or ten groups. These are the 'strata' for the survey:

within each, a number (say ten) of 10-km squares would then be chosen at random and an appropriate number of air pollution sampling points would be laid down also at random, in each. The results would not only say something about the variation in pollution over the country, but something of the variation *within* and *between* urban and rural categories. If it was then decided that urban sites were of greatest concern, the two strata of highest urbanisation in the survey might be emphasised by choosing twice as many sample squares within them as in the strata of lowest urbanisation. In practical terms the number of sample points, the number of strata, and the number of sample squares in each stratum should not be pre-judged – the initial stages of the survey itself would indicate the degree of variation, and taken together with the resources available for the study, could lead to a decision on a sampling pattern that would have the greatest sensitivity and resolution obtainable with these resources. But the point to emphasise is that an objective approach of this kind to environmental factor surveillance and monitoring, whether in the air or water, is most desirable. In the real world it is not easy to achieve. Not all randomly chosen sites are accessible, safe from vandalism, provided with power or free from intensive and very local pollution. But even these difficulties need not prevent substantial objectivity of approach.

A similar objective approach is needed to the pattern of sampling in time. As an initial step the variation in levels over time should be determined either by continuous measurement or frequent sampling, but how intensively this is done must depend on judgements about the resolution needed. For example, if the aim of monitoring is simply to check compliance with the WHO long-term goal for smoke and sulphur dioxide defined in terms of annual mean concentrations, then a sampling system which records monthly average figures may suffice: this will reveal the annual pattern of variation and the period when any problems are liable to arise. For the short-term goal, set in terms of 24-hour means, clearly nothing less than 24-hour values will suffice and it may be necessary to sample, at least at peak periods, hourly so that the diurnal fluctuation that causes high peaks is detected. This is obviously necessary when working to a sulphur dioxide primary standard like that in the United States in which it is stipulated that emissions must be below the limit for 95 % of the time. For ozone, liable to damage vegetation if levels are high for a matter of hours, continuous sampling may be needed under circumstances when there is a high probability of a peak. By analogy with the concentration of sampling of pollutants in urban zones, intensification of sampling frequency may be introduced once concentrations exceed a 'warning level'. Generally, sampling needs to be an order of magnitude more frequent, or cover an order of magnitude less period of time, than the periods to which a standard or goal relates. The costs, naturally, rise sharply with sampling intensity and provide a deterrent to excessive elaboration.

7.2.4 *Some characteristics of target monitoring schemes*

Target monitoring involves a comparable approach. Those types concerned to record changes in the distribution of ecosystems or of species, or changes in species performance over wide areas, are often best catered for by the periodic repetition of a thorough environmental survey. Such surveys are of three general kinds:

(*a*) within site surveys designed to define the degree of variation in a particular area (which may be of any size);

(*b*) between site surveys designed to record how a particular system varies from site to site;

(*c*) within species surveys designed to define variation in a species over its range.

Any such survey can become a monitoring programme by suitable repetition. In such instances the first survey is sometimes spoken of as a 'baseline survey' because it defines the starting point, and may be more extensive in scale than later surveys which may concentrate on components of the system: they can be placed in a context by the existence of the baseline survey, so long as the methods are comparable.

These kinds of survey, like the surveillance of factors, demand an objective approach. This has not always been adopted. In any activity of this kind, the usual aim is to describe the environment in terms of certain selected parameters that are capable of being mapped with the conventions currently available, mainly lines and variously shaped dots on paper. Hence continuously varying slopes are abstracted as contour lines, water and land as colour contrasts with the interface as a border picked out in conventional signs that denote its sandy, shingly or rocky nature, and ecological systems as crude symbols depicting woodland, orchard, moorland, wetland or (by absence of colour) farmland.

There are two ways of enhancing the detail in mapping environmental systems such as vegetation (taken as an example). One, most commonly used, is the 'obvious' method – that is the one that intuitively appeals to people by its apparent simplicity. In it, one takes a look at the region and decides on the categories that can be used to define its vegetation. The constraints are that these categories should have mappable – that is linear – boundaries which need to be distinct, and that the categories together must give complete coverage of the area. The result is to take conspicuous dominant plants like trees, shrubs, heaths, rushes or grasses and make up categories such as oakwood, heather moor, marsh, fen or mat-grass and deer-sedge moorland. The trouble with this is that it obscures the real intergradation and heterogeneity that the environment often displays: linear boundaries rarely exist between vegetation types unless these chance to coincide with a topographic or geological feature. It also depends heavily on the 'judgement' of the observer, and its repeatability may be low if another observer tries to re-survey the area after a decade or more. The subjectivity of the approach causes endless doubt over whether apparent change between two surveys by different people twenty years apart is genuine, or whether one called something 'heather moor' which another called 'boggy moorland with heather'. The other trouble is that this approach is structured to produce an average rather than record the full degree of variation in the environment.

The alternative and more objective approach is to begin by a desk study to define the field of variation in habitat. This, for example, might choose every tenth 1-km grid square on the map, or might locate a sufficient series of sample plots at randomly chosen points. For each sample point or square a range of features that define the main characteristics of the area as a habitat would be recorded. These might include altitude, slope aspect and steepness, proximity to coasts, drainage systems and presence of or proximity to built-up areas and communication axes. This information can be sorted so that the sample sites are arranged along the chief axes of variation, then divided into groups which show greatest affinity. This provides a set of habitat categories which can be used to stratify vegetation sampling – taking randomly chosen squares or sites in each category and within them recording in the field all the plants present, their relative abundance and a number of associated features (including soil) at sample points chosen randomly in each square or site unit. This detailed information is used to examine the axes of variation in the vegetation and to define categories which bring together the samples that are most alike. The procedure is thus a stratified random survey like that outlined for surveying the pattern of variation in an environmental factor. In a similar way, the system records the degree of variation in the vegetation and environment in the region studied and provides rigorously derived groupings which do not depend on the value judgements of the observer. The intensity of sampling clearly determines how completely that variation is recorded, but it is possible to evaluate this by statistical means and settle for a sampling intensity that will record the variability to the desired level of resolution.

Repetition of the whole survey procedure would then provide an objective means of evaluating environmental change – change in mappable habitat parameters like urbanisation and in plant species distribution which might be correlated with the changes. The same sample sites might be re-visited, to provide exact bases for comparison. Selected plant groups might be chosen for monitoring, recording lichens, for example, in a scheme directed to air pollution, because of the known sensitivity of these plants or selected shore animals known to vary in performance with oil pollution in a study of refinery effluents. Animals known to accumulate residues of particular pollutants could be collected from sites which had earlier been related to the national dimensions of variation in habitat or in emissions of those substances. The important thing is that the sample can be placed within a known overall pattern of variation and protected from observer bias: it is also important that the frequency of repetition is sufficient to give warning of changes on the time scales sought.

Just as the effects of pollution described in Chapter 5 are manifest at many levels, so biological target monitoring can involve:

(a) measurements of biochemical changes within the tissues of organisms;
(b) measurement of physiological and performance changes of individual organisms of selected species;
(c) measurement of population in selected species;
(d) measurement of the distribution of selected species;
(e) measurement of changes in the functions (or performance) of selected samples of ecosystems;
(f) measurement of changes in the distribution of ecosystems.

As Chapter 5 has shown, the effects of a habitat variable often build up hierarchically through several of these levels, overcoming or 'saturating' an actual homeostatic mechanism or its analogue at each level. It follows that selection of the biological parameter to measure is critical in relation to the use to be made of the information, and our concern to protect particular targets. If our concern is simply to detect biological changes that may alert us to changing environmental factors, monitoring of the distribution or performance of ecosystems, populations, individual organisms or biochemical systems within them may all be apposite and the need is for selection of good indicator organisms, systems or substances, based on an understanding of criteria. Species that accumulate substances present at low (and often undetectable) concentrations in the environment (e.g. metals in shellfish, lead in *Ranunculus fluitans*, PCBs and organochlorine pesticides in bird fat and eggs) are particularly suitable for the monitoring of these substances as factors in the general environment. So (as a variant on the same theme) are biological materials that have a physical capacity to trap pollutants: the use of bags of dried moss as collectors of air pollution is well known in this context (e.g. Welsh Office, 1975). If, on the other hand our concern is to protect the stability of ecosystems we shall need to step back and monitor either performance change in those systems manifest well before the point of unacceptable change, or changes in the presence or absence of species whose loss does not threaten the whole (e.g. as the loss of corticolous lichens does not threaten the stability of an oak wood), or changes in the performance or populations of species of known sensitivity and ecological importance. Similarly, to gain advance warning of population effects in a species, it is necessary to monitor chemical or performance variables that are conspicuous well short of the threshold of population changes (as when eggshell thickness is monitored in birds).

The general rule, therefore, is this. If one is embarking on a monitoring exercise in order to protect a target (whether or not this is the organism monitored), it is essential to build in an adequate 'lead time'. It is no good waiting until the species concerned has collapsed, or areas of land have gone out of production, or thousands of people have been admitted to hospital. It follows that:

(a) if biological target monitoring involves direct observation of the species which it is desired to safeguard, we must observe some feature of its physiology or performance or some residue level within it that shows detectable change well before the threshold of unacceptable damage.

(b) if biological monitoring is of a species known to be more sensitive than the species it is desired to protect, the difference in sensitivity must be enough to allow intervention after the 'indicator' organism has shown adverse effects.

7.3 Monitoring in operation

7.3.1 Global, regional and national surveillance and monitoring

It is obvious that surveillance and monitoring schemes differ greatly in their extent in space and in time – from an area of a few square metres within or around an emission, and a frequency of observation that is so rapid as to be virtually continuous, to the area of the whole planet, and a periodicity that seeks to define no more than year-to-year trends in a variable like the carbon dioxide content of the atmosphere. If one was being strict about it, restricting the term monitoring to measurements in relation to a standard or control, most global programmes are of surveillance. The only world-wide pollution controls are on the dumping of wastes in the ocean, under the London Convention (1972) and on oil release from ships under IMCO Conventions, and the monitoring of oceanic pollutants organised by IMCO (the Intergovernmental Maritime Consultative Organisation) as the body administering those Conventions was accordingly the only world-wide scheme of monitoring in the narrow sense operational in 1975. But many international agencies (such as UNEP, WMO, IOC and WHO) look to global schemes of environmental measurement to guide their member governments towards harmonised action to combat the pollution of the world environment.

Global, regional or national schemes all link observations from sample points to build up a general picture. Often the linkage is in several stages. National schemes commonly link local networks of sampling points designed primarily to give a river management authority or a municipality information about the pattern of contamination in the area under its charge. International schemes generally then link the national networks into a larger system rather than operate field measurements directly and sometimes select only a few of the national measurement stations available. However, international schemes like the Global Atmospheric Research Programme (GARP) of the World Meteorological Organisation (Steward, 1973) have often led to the supplementation of national schemes with additional stations or additional sampling procedures.

Munn (1973) (on behalf of SCOPE) has drawn up proposals for monitoring at the global level. He stresses that for designing such systems or predicting trends based on their results, there must be a conceptual model which describes the system the monitoring scheme will sample. The measurements must be put into context. Biogeochemical flux diagrams like those for carbon and sulphur in Figures 9 and 19 can be made into models of this kind, stating the quantities of a substance present at the starting point of the scheme in the 'compartments' of environment (represented by soil, vegetation, fresh-water run-off, lakes, surface seas, marine sediments or various air layers, regionally or globally), and the rates of movement between them. The monitoring schemes can then be designed to check on key components of such models or the rates of key processes. A pathway diagram like Figure 50 can equally be converted into a model for such a purpose, inserting the emissions from a known source in a known time, the amounts of pollutant at points along the pathway when the scheme starts, and the amounts reaching chosen targets whose responses are to be watched. This kind of thing is more usually done implicitly than explicitly by those designing monitoring and surveillance schemes or research projects along comparable lines (that is to say, most of them think about the systems they are sampling, but few submit to the discipline of actually writing the model down and putting figures into it!). Ideally, the models should be set out and published, alongside the design of the sampling system and predictions of the expected sensitivity of the scheme, at the outset, so that the model is itself tested as a main element in the whole programme.

At the global level, surveillance and monitoring are especially concerned with major geochemical cycles. As SCOPE (1976) points out, monitoring these 'can draw attention to any significant alteration in the amount of substance in circulation and any appreciable changes in the pattern of the cycle itself which have been caused by man's activities or by a significant shift in natural events'. Often knowledge of dose–response relationships is used to choose factors known to affect the functioning of physical or biological systems – and living targets known to accumulate pollutants or be especially sensitive to their effects. It is clearly important to cover the global field of variation if a meaningful world picture is to be presented, and Munn (1973) recognises this when insisting on the need for remote, or baseline, stations away from known human interference (see also SCOPE, 1971; WMO, 1971, 1974) as well as regional stations in intermediate positions and others in sites known to suffer a high level of human impact. The choice of actual sites has not, however, followed a strictly logical examination of world environmental variation and the results may accordingly be difficult to interpret as a strict quantitative indication of world patterns and trends.

Table 23 lists the substances and 'environmental stress indicators' recommended by Munn for coverage in Stage I of UNEP's Global Environmental Monitoring System (GEMS), as well as variables considered more appropriate for local surveillance. Intercomparability of methods, intercalibration of equipment (and measures to ensure thorough training of the people involved), careful checks on results and effective data handling using comparable methods are all vital to the success of such schemes. The United Nations Environment Programme is organising the first stages of GEMS by linking existing schemes such as GARP or pilot marine monitoring programmes into an emerging global framework.

There are also regional monitoring schemes. For example, the International Council for the Exploration of the Sea (ICES), the Commission administering the Oslo Convention (Command 4984, 1972), and the group of countries bordering the Baltic, have all agreed on monitoring programmes for the areas of water with which they are concerned: the Mediterranean is likely to be covered by a parallel scheme following agreement in 1976 on measures for reducing its pollution. In Europe, the Commission of the European Economic Community is developing plans for harmonised environmental monitoring, which may be linked to the emerging quality objectives or standards under the Community's environment programme. In North America, bilateral programmes between the United States and Canada to control pollution of the Great Lakes have been linked with agreed monitoring. As at the global level, many of these regional activities link together national schemes which by themselves are somewhat fragmentary.

7.3.2 *Monitoring in Britain*

Britain provides a reasonably representative example of an industrial country with a fairly extensive monitoring system. Monitoring of pollution in Britain has been reviewed by the Central Unit on Environmental Pollution (DOE, 1974). This review recorded 100 separate programmes in

Table 23. *Priority pollutants and environmental stress indicators proposed for monitoring*

Substance or indicator	Medium
(a) Recommended for inclusion in GEMS phase I	
Airborne sulphur dioxide and sulphides	air
Suspended particulate matter	air, water
Carbon monoxide	air
Carbon dioxide	air, ocean
Oxides of nitrogen	air
Ozone, photochemical oxidants and hydrocarbons	air
Toxic metals	
(i) mercury	man, soil, food, water, biota
(ii) lead	man, soil, food, water, biota, air
(iii) cadmium	man, food, water, biota
Halogenated organic compounds, especially DDT and metabolites, dieldrin, PCBs	man, soil, food, biota
Petroleum hydrocarbons	water
Selected indicators of water quality	
(i) biological oxygen demand	water
(ii) dissolved oxygen	water
(iii) pH	water
(iv) *E. coli*	water
(v) ammonia	water
Nitrates, nitrites and nitrosamines	water, food, soil
Specific radionuclides (cadmium-137, strontium-90)	all media
(b) Recommended for local or regional monitoring where problems exist	
Soluble salts of alkali metals and alkaline earth metals	groundwater, soil
Eutrophicators (nitrate, phosphate)	water
Other substances known to have caused local problems, e.g. arsenic, boron, phosphorus, selenium, fluoride, heavy metals	soil, water, food
Noise	air
Waste heat	air, water
Ammonia	soil
(c) Not recommended for phase I because feasibility of systematic monitoring has not been established	
Polycyclic aromatic hydrocarbons	
Asbestos	
Allergens	
Selected microbial contaminants	
Mycotoxins	

From Munn (1973).

Table 24. *Summary of monitoring schemes in Britain listed by the Department of the Environment in 1974*

Sector	Measurement	Number of schemes[a]
	(1) *Factor monitoring*	
Air	Composition of air masses: national or local	16
	Composition of air around individual industrial sources	5[b]
	Pollution of the air by road traffic	2
	Deposition of pollutants in rainfall	9
	Dry deposition of pollutants	3
Fresh water	Composition of fresh-water systems	9
	Inputs and sinks	6
	Residue levels in target organisms	4
Land	Deposition of wastes	1
	Residue levels in soil	4
	Residue levels in organisms	10
	Residue levels in foods	4
Marine	Composition of seas and estuaries	10
	Inputs and sinks	6
	Levels of pollutants in organisms	9
	(2) *Target monitoring*[c]	
Air	Corrosion of zinc by atmospheric pollution	1
	Effect of airborne fluoride on lichens and other plants	1
Fresh water	Use of *Bifidobacterium* as an indicator of faecal pollution	1
	Micro-organisms in Thames and Kennet	1
	Productivity in Loch Leven	1
Land	Pesticides and mercury as factors in wildlife deaths	2
	Effects of metal-contaminated sludges on plant growth	1
	Effects of slurry disposal on grassland	1
Marine	Effects of new sewage outfall on benthic ecosystem	1
	Weights of landings of fish and shellfish to assess possible effects of pollution	1
	Measurement of toxic metals in seawater and biological indicator systems	1
	Monitoring *Escherichia coli* from sewage effluents by use of cleansed mussels	1
	Effects of environment and contaminants in transport and exchange processes in marine invertebrates	1
	Marine pollution effects on meiofauna	1
	Size and output of nesting colonies of seabirds variously exposed to pollution	1

For notes see opposite page.

some way concerned with monitoring or surveillance. These are sum-marised in Table 24. It is clear that factor monitoring in the environment predominates: process and emission monitoring by individual firms, however, was excluded from the survey and is probably the largest single category. The Alkali and Clean Air Inspectorate, Industrial Pollution Inspectorate in Scotland, and regional Water Authorities and River Pollution Authorities were the main official agencies with process and emission monitoring programmes. It was also clear that many of the studies listed in the review were more research than action-orientated in character. The number of surveys actually set up to monitor a contaminant in air, water or soil as an aid to environmental protection was small. There were relatively few target monitoring schemes (listed in full in Table 24) although the report did record rather extensive corrosion surveys by the CEGB and drew attention to the potential value of epidemiological and biological monitoring schemes. It is also evident that the different schemes have been developed separately, in response to different needs, with little thought about their inter-relationships (a defect that is now being remedied by the Monitoring Management Groups).

The largest environmental pollution monitoring network is the National Survey of Smoke and Sulphur Dioxide, operated by the Warren Spring Laboratory at Stevenage (Department of Industry) (Warren Spring, 1972). This survey uses about 1200 sites in Great Britain. It depends on local authorities and industry (e.g. the British Steel Corporation and the Central Electricity Generating Board) to run the sites and collect the samples, which are sent to Stevenage for evaluation. Not surprisingly, the sampling pattern is partly dependent on the location of willing operators. It is mainly concerned with urban areas, where levels of target exposure are highest. This is understandable when it is remembered that the origin of the investigation was the London smog of 1952, and its aim

---

The individual schemes vary greatly in size, and some have been scored under two headings to reflect their extension over several categories. The programme is continuing to expand, under the guidance of the Monitoring Management Groups referred to on p. 188.

[a]   Schemes are listed here without regard to their magnitude, and they vary greatly in size and extent. The 16 studies of air composition range, for example, from the over 1200 sites at which smoke and sulphur dioxide are measured to several single-site studies at MAFF Experimental Farms.

[b]   The CUEP Report confined itself to Government supported schemes. Many monitoring programmes around factories are operated by industry, and not listed.

[c]   The target monitoring schemes are listed in greater detail, but this list, too, is incomplete: it evidently overlaps substantially into research, and could be extended to encompass numerous University projects as well as schemes of biological surveillance not specifically linked to pollution, but providing general indications of environmental quality.

Table 25. *Main atmospheric components monitored at several sites in Britain in 1976* (ad hoc *studies around factories and roadsides omitted*)

Substance	Sites	Basis of selection
Smoke ⎫ Sulphur dioxide ⎬	About 1200[a]	Volunteer local authorities: no systematic randomisation or stratification
Grit and dust	Several hundred	Local authority initiative
Airborne metals	20	Choice from sites offered by local authorities, to give fair spread of samples
Acidic particles and aerosols	20	Choice from sites offered by local authorities, to give fair spread of samples
Nitrogen oxides, oxidants, hydrocarbons	3	Pilot study: convenience in operation

[a] About 600 sites only provide valid sequences of figures over 2 years.

was to pinpoint areas where smoke and sulphur dioxide levels were unacceptably high so that clean air legislation could be brought to bear. The sampling was therefore deliberately stratified to concentrate on urban zones, but it may be criticised because it was not preceded by a preliminary random survey of the national range of variation – or that within urban 'hot spots'. To be fair, however, a rigorous sampling system that was not related to volunteer operators would be much more costly, and it may be questioned whether the increased precision would be worth this extra cost.

Table 25 lists the parameters now monitored in the British air, and the number of sites in each case. The design of the system is being reviewed (1976) by the Department of the Environment. There is no doubt that the scheme has proved valuable in documenting smoke and sulphur dioxide levels at many points in Britain and in demonstrating for example, that in a number of urban areas the WHO short-term objective of 250 μg/m^3 smoke in the presence of 500 μg SO$_2$/m^3 is exceeded on some tens of days in the year, and hence that clean air legislation still has some way to go.

In fresh water, monitoring has been designed to sample all major discharges and also to record the composition of British rivers at some 175–225 points, mostly just above tidal limits or at the confluence of major tributaries. These sites are being selected by Water Authorities (or River Purification Authorities in Scotland). The water analyses commonly carried out in routine surveys by Water Authorities are set out in Table 26. The siting of sample points has so far been non-random, the rivers and tributaries being chosen by size, volume of flow, or known existence of

Table 26. *List of water analyses commonly carried out by British water authorities*

Temperature	Albuminoid/organic nitrogen
Suspended solids	Ammoniacal nitrogen
Transparency	Nitrite nitrogen
Electrical conductivity	Nitrate nitrogen
Chlorides	Phosphate
Alkalinity (as $CaCO_3$)	Sulphate
Hardness (Ca, Mg)	Metals
pH	Cadmium
Dissolved oxygen	Chromium
Biochemical oxygen demand	Copper
Permanganate value	Nickel
Dichromate value (COD)	Zinc
Phenols	Cyanides
Oils	Detergents

From Hawkes (1974).

pollutant problems. The results will be valuable in local terms, like the air pollution surveys, and may allow general statements about overall patterns and trends in British rivers because they can be put in context by a recent national survey in which all river reaches were assigned to one of four pollution categories (DOE, 1971, 1972).

This rivers survey could be used to stratify a more objective sampling scheme, laying down reference points on an appropriate number of randomly selected reaches in each pollution category. Such an approach would document the national pattern of conditions, but would not do all that the present system does in giving specific information about individual rivers with known problems. Being located at tidal limits where the combined drainage of a basin reaches the sea, the present chain of sample points moreover gives an indication of the inputs to the oceans – or more exactly estuaries – which is in itself a statistic worth having.

British schemes of biological target monitoring, and the underlying philosophy, have recently been reviewed by NERC (1977b). These schemes are more fragmentary. As Table 24 indicates, the largest number of schemes concern the recording of residue levels, especially of metals or pesticides in wildlife. Some of these schemes depend on analysis of carcases found dead in the environment and collected for laboratory study. The occurrence of deaths among wildlife on a substantial scale has in the past proved valuable in alerting us to unexpected environmental problems (as mortality among predatory birds and mammals alerted us to the unexpected side-effects of DDT and related pesticides while the mass death of seabirds in the Irish Sea in 1969 pointed to the potential hazards of

PCBs). Unfortunately, animals found dead in this way are not an unbiased sample of the populations they came from. They contain an abnormally high proportion of aged animals and those with high levels of toxic chemicals. It would be much preferable in scientific terms to sample randomly, killing animals specifically for monitoring. Not surprisingly, since many of the most valuable species as accumulators of residues are relatively rare bird and mammal predators, such a procedure is not acceptable on conservation grounds. But if certain common or 'pest' species – like starlings, gulls or crows – could be chosen the result would be a system that is more readily interpretable, as random sampling of marine or fresh-water organisms is.

As Section 3.2.4 stressed, food is another important pathway by which pollutants reach man: for some, like lead, the most important. Moreover there are some potentially very hazardous materials, like mycotoxins, that can contaminate food. The analytical surveys of food carried out in Britain by the Ministry of Agriculture, Fisheries and Food and the Laboratory of the Government Chemist have been reviewed by Egan and Hubbard (1975). The sampling is based on purchase of the components of a representative diet in twenty-one towns distributed over the country (the total diet study), backed by selective studies on individual foodstuffs which may present special problems (as, for example, fish does for methyl mercury).

Organochlorine pesticide residues have been monitored in animal fat since the 1960s, and polychlorinated biphenyls have been kept under surveillance in later years. Among metals, mercury, cadmium and lead have been the subject of special studies, while arsenic and selenium have received some attention and copper, chromium, cobalt, zinc and tin are now being examined. Searches have been made for mycotoxins, nitrosamines and polynuclear aromatic hydrocarbons, none being found in significant quantities. Such monitoring of food is an essential counterpart to the monitoring of water supplies and of air, water and targets in the general environment.

Many schemes of biological monitoring are not directly concerned with pollution. The most numerous involve the recording of the presence of species in the squares of the National Grid (usually 10-km squares) by the Biological Records Centre of the Institute of Terrestrial Ecology (NERC). Such surveys, repeated periodically, allow changes in the distribution of species throughout Britain to be followed. Such records do not in themselves allow the causes of changes to be ascertained, but a good deal of associated information about the habitats of rarer species, the size of their populations and alterations in the environment that may account for the loss of a record are often obtained. On the other hand, the value of distribution surveys of sensitive plants (like lichens) known to respond to pollutant levels has already been stressed. Lichens can be transported, on an appropriate stone or disc of bark, from an unpolluted to polluted area and used as 'instruments'. This has been done to examine the spread of effects associated with an aluminium smelter emitting fluoride in North Wales (Perkins, 1975). A similar method has been used by German scientists in the Ruhr (Schonbeck and Van Hauf, 1971). Such 'presence and absence' surveys display changes especially in the rarer and more sensitive species. They are not likely to demonstrate more subtle performance variations or population variations in common species. Many of these are probably common because of their resilience in the face of environmental stress. Yet, because of this capacity, it may be just these species that can display performance changes that can be calibrated against environmental variables. This is known to be the case in certain marine algae (*Ulva, Laminaria*) (Burrows, 1971), common shore animals (dog whelk, *Nucella lapillus*, barnacles, *Balanus* sp.), some fresh-water animals and certain birds. E. B. Cowell and W. Syrett (personal communication) have used variations in the performance of the supralittoral lichen *Verrucaria* in monitoring the effect of pollution from an oil refinery in Norway. Baker (1976b), Dicks (1976), Cowell (1971) and others have shown how the distribution and performance of salt-marsh plants and intertidal and sublittoral animals varies in areas subject to chronic pollution from refinery effluents at Milford Haven and in Southampton Water. Biological indices (Woodiwiss, 1964; Hawkes, 1974; Hellawell, 1976), of which that developed for the Trent is best known, similarly seek to classify the degree of fresh-water pollution by observing the abundance of selected river organisms – especially widely distributed groups. So far these kinds of performance variation have been used only in local surveys, often to evaluate the effects of single industrial developments.

Similar performance variables in birds have proved valuable in wider programmes of surveillance and monitoring. For example, at the height of organochlorine pesticide usage in Britain there were changes in reproductive success and higher adult mortality in herons, and at the same period predatory birds such as eagles, peregrine falcons and sparrow hawks showed marked changes in population size and distribution. The latter would have been shown up by surveys based on presence and absence: the former would not. Yet research established the particular value of the heron as an indicator species, because it remained common and widespread yet showed clear performance variation that could be correlated with this particular type of pollution.

A third approach is through reference sites, as a nationwide network, at which the flora and fauna would be periodically surveyed (in terms of population size rather than just presence or absence). It has been suggested that national nature reserves could form such a series of reference sites. Such reserves, however, do not necessarily sample the full range of variation in British ecological systems (the accent is often on species diversity, rarity and extremes in the ranges of variation rather than on the transitional and relatively less diverse systems that are widespread in much of the country). Moreover, the distribution of the reserves is negatively correlated with that of the main impact of urbanisation from which the threats are likely to come. It would probably be better, therefore, to select reference sites from an initial stratified-random survey of the range of variation in British environmental systems. Even at such a network it would be an extremely heavy task to identify all the species present and provide reliable estimates of their abundance. It has not been established that so all-embracing an approach is necessary. Just as within taxonomic groups certain especially valuable indicator species may be selected, so at these reference sites it would probably be best to concentrate effort on an intensive assessment of the abundance and performance of such indicators (e.g. lichens, vascular plants and selected invertebrate groups).

Neither factor nor target monitoring is adequate alone, and the sampling systems need to be integrated so that correlations can be established and causes deduced and tested by research. To do this on a regional or national scale, the information must be produced in a suitable common format. It is no good having one programme, however objective, sampling on a basis of administrative districts and another on grid squares, while a third can only provide statements about the average levels of a pollutant in a whole river catchment. For example, health statistics in Britain are collected by administrative district, which is not a very helpful unit for scientific studies because districts vary in size and do not provide strictly comparable samples. Nonetheless, it would be possible to use districts as units within which to sample atmospheric or river-borne pollution and metal or pesticide residue levels in organisms or soil. Surveys of target and factor could then be linked – but only if the data were reduced to the common sample unit. The need is for co-ordination at the design stage in a way that takes account of the uses to be made of the data, and at the interpretative stage so that the value of each programme in reinforcing or testing the next is not lost.

Following publication of the report on the monitoring of the environment in Great Britain (DOE, 1974a), in which the need for such co-ordination was stressed, six monitoring management groups have been established by central government. Four of these are concerned with sectors of the environment (air, land, fresh water and sea) and two with 'horizontal' themes (the human and ecological aspects of environmental health). All the official interests with responsibilities in these areas are represented on the groups which are co-ordinated by the Central Unit on Environmental

Pollution in DOE as the national focal point for monitoring. CUEP is engaged in planning the across the board co-ordination of the monitoring and assessment of pollution, as part of the analysis of change in the national environment as a whole.

7.4 *The components of a monitoring scheme*

In principle, it should be possible to devise a system of monitoring which defines major environmental and target trends that, like economic indicators, provide a measure of the 'environmental health' of a region and its potential targets. Such a system should encompass:

(*a*) exposure or internal monitoring of all targets where no excess exposure is tolerable and a 100% sample is essential;

(*b*) environmental monitoring on a statistically meaningful plan, where known contaminants are being emitted and there is a known relationship between their levels and risk or effect;

(*c*) target monitoring where there is a reason for guarding against changes in the status, performance or production of the targets concerned, this monitoring being designed to give early warning of forthcoming effect; it may involve observation of a sample of the target whose protection is desired or of other more susceptible targets;

(*d*) target monitoring where the targets concerned accumulate hazardous materials and may be sampled according to a statistical plan which allows residue levels to be related to factors in the environment.

8. Costs and controls

8.1 *Basic questions*

The imposition of controls on pollution depends on an evaluation of:

(*a*) the need for such action, which in turn depends on assessment of the damage being caused, the value placed upon the targets, the risk if no action is taken, and the benefits to be gained from actions, and;

(*b*) the nature of the action required, which must in turn demand assessment of the techniques available, of the costs a particular level of abatement will require, and of the administrative and legal machinery to hand.

Reduced to the simplest financial terms, the critical questions are:

What are the present damage costs?
What are the predicted damage costs if present policies remain?
What degree of abatement will reduce these damage costs to acceptable levels?
What are the control costs for this degree of abatement using present technology and under present administrative machinery?
Who pays, and who benefits?

and perhaps

What new technology and new organisation can be developed to give the degree of abatement required at less cost?
What are the predicted costs and benefits of forseeable future options?
Who would pay, and who would benefit?

Figure 51 sets out the general relationship. If there is no pollution abatement, damage costs are clearly maximal, and control costs zero. If there is 100% abatement the opposite situation prevails. In most situations an intermediate is sought at which the net cost to the community is minimal (i.e. the lowest point on a curve which is the sum of the two in the figure) and this, in the figure, is at point X. This is clearly the degree of abatement which should be sought in most cases.

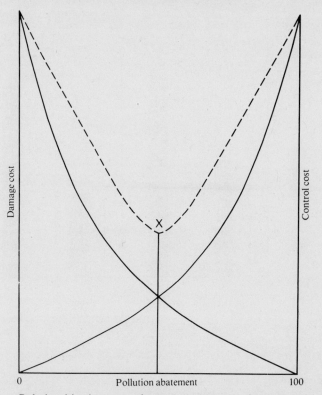

Figure 51. Relationship between damage costs, control costs and pollution abatement (arbitrary units). The dotted line is the sum of the two curves and indicates net cost. Note that if damage estimates are increased or if control costs are lowered, for example by new technology, the optimum shifts towards higher abatement: the opposite happens if the damage is shown to be less than feared or if control costs rise.

8.2 *The costs of damage and the evaluation of benefits*

Such an analysis is theoretically unchallengeable, except on the argument that there are damages attributable to pollution, and benefits from the environment, that simply cannot be put into money terms (this argument is often deployed when landscapes or historic monuments are threatened with irreversible change because of proposed new development). Others will hold that there is no theoretical reason why such values should not be costed – either as a single figure, representing a consensus of social evaluation at the time, or as a range, reflecting the diversity of views in the community. All figures on both sides of the equation are bound to be ephemeral, after all. The value of targets affected by pollution – and the benefits of the pollution-generating activity – certainly can be expected to

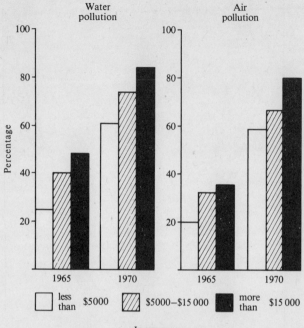

Figure 52. Environmental concern by income level. Responses to the question 'Compared to other parts of the country, how serious in your opinion do you think the problem of (air or water) pollution is in this area – very serious, somewhat serious, or not serious?' The figure gives the percentage in each of three income groups responding 'very serious' or 'somewhat serious'. Source, Council on Environmental Quality (CEQ, 1973) based on data from the Opinion Research Corporation, USA.

be revised as understanding of the effects of particular doses of pollution increases. These values also reflect social attitudes and standards of living: an affluent society commonly sets more store by wildlife and natural beauty (Figure 52). Control costs are also affected by innovation, generating new and cheaper control methods, and by changes in industrial costs which may make control more expensive. The system is a dynamic one, and the point of balance can and will shift in both directions over time.

8.2.1 *Damage costs*

What information about damage costs exists? In 1952 Bleasdale (1959) calculated that agriculture in south-east Lancashire was losing £2.6 million per annum through air pollution with smoke and sulphur dixide. The Beaver Committee (1954) put the cost of air pollution in Britain at £250 million per annum. More recently the Programmes Analysis Unit (PAU,

Table 27. *Estimated cost of damage due to air pollution in the United States in 1968 and projected costs in 1977 assuming no improvement in control*

Type of cost	1968 (billions of dollars)	1977 (billions of dollars)
Damage to human health	6.1	9.3
Damage to materials and vegetation[a]	4.9	7.6
Reduced property values	5.2	8.0
Totals[b]	16.2[c]	24.9[c]

From CEQ (1973).
[a] Damage to crops assessed at 0.1 billion dollars in 1968; damage to materials based on studies of 50 substances believed most susceptible to pollution.
[b] Damage to animal health and costs of cleaning soiled materials not estimated.
[c] The 1968 figure is in 1968 dollars while that for 1977 is in 1970 dollars: there is consequently some inflation of the 1977 figures.

1972) set the cost at around £400 million per annum (but with a huge margin of error). They estimated that two-thirds of the damage was due to domestic fires, that industry and electric power generation cost £120 million per annum and that motor vehicle pollution cost £35 million per annum. In 1968 the United States Council on Environmental Quality put their annual cost of air pollution at around $16 200 million (Table 27; CEQ, 1973), and calculated that without improved pollution control these costs would mount to $24 900 million per annum by 1977.

The evaluation of the costs of damage to human health, or curtailment of life, is particularly intractable and the subject of continuing debate. Should one cost loss of life in terms of a standard compensation to next of kin? Does one capitalise the sum required to sustain mean earnings over the remaining life span of each individual? Does one say that a shortening of mean expectation of life of the whole population of a country by one day should be valued likewise in terms of average income – which, taking Britain in 1973, meant that with a national personal income of around £75 billion per annum, or £200 million a day, a day in a persons life would be valued at about £4? Most people would pillory such a calculation as nonsense. The PAU assessed 'economic GNP costs incurred by the nation' (loss of output, costs of treatment) 'economic costs not in GNP' (e.g. incurred by housewives, students or uninsured working women) and social costs, e.g. disamenity of illness and 'social cost of premature death' (which worked out at £1000 per year of life saved). Another approach is to calculate people's own value for their lives by observing their behaviour – for example assessing how much increased income is demanded for hazardous employment. As Chapter 7 pointed out, the lack of agreement about degrees of risk and consequent action is reflected in the wide variation in expenditure to avoid hazard in different industries.

Obviously, this is a very difficult area. It is also difficult to balance this kind of cost against agricultural or forestry costs (much easier to assess if actual crop losses can be quantified either in terms of diminished yield or yield foregone because pollution prevents the growing of the most productive species). It is similarly possible to cost the impairment of the quality of other components of the environment. For example, lost river quality can be assessed in terms of the value (actual or potential) of fisheries (e.g. trout fisheries in Lakes Superior, Huron and Michigan declined from around 4, 5 and 6 million lb/year before 1920 to 0·5, 0 and 0 respectively in 1965, largely due to the invasion of the lakes by sea lampreys, and other fisheries also suffered dramatic declines (Smith, 1971; Beeton, 1971).

The point about such analyses is that, even if the estimates are imperfect, they do bring different elements in the system to a common basis of financial assessment. Like may not be being compared strictly with like – but that is one of the objects of the whole approach: to find an acceptable way of comparing unlikes in a way that leads on to coherent policies. We are not, however, at the point where damage costing is so far advanced that this approach can be the dominant influence on policy.

Very often, the standards that have been set have not emerged from a costing exercise. The World Health Organisation short-term objectives for smoke and sulphur dioxide in urban air, for example, are set at the threshold above which hospital admissions and deaths of patients most vulnerable to emphysema and bronchitis are known to have risen in London smogs. It could be argued that the standard implies that a significant rise in mortality or hospitalisation from this cause is a cost that the community cannot bear, whereas damage costs are acceptable below this point (both arguments could be challenged). Rather similarly, recent controversies over the emission of lead to the environment from vehicles, which has not been shown to cause any deaths (in contrast to the continuing slight mortality in Britain and considerable mortality in America among children who chew old paint), may indicate a social judgement that the elevation of blood lead to say 30–40 μg/100 ml, with possible risk to the most vulnerable in the community, is a potential cost that cannot be accepted. But there are no reliable figures defining what those costs are – and still less has there been any balancing of them against other social costs stemming from other activities.

8.2.2 Environmental values

A damage cost is a demonstrable loss related, in the present context, to a defined level of pollution. It should generally be something that can be measured because it has happened. It may well differ from the value set upon a component of the environment, advanced in an argument about the desirability of potential development. For example, if the yield of a salmon river was 200 fish a year, of average weight 15 lb, marketing at £2 per lb, the damage expected from a development that would make the river incapable of supporting salmon but in no other way impaired it might be assessed at £6000 per annum. The sporting revenues, derived from fishing licences, might add substantially more – maybe £100 000 – but the maximal annual damage cost of the impairment would be unlikely to exceed £200 000. The loss in the value of the river as a salmon fishery might however also be estimated by considering the diminution in market value of the fishing rights to be equal to the annual sum which all those concerned would be willing to pay to prevent its loss in quality, or to restore it, should it deteriorate. This might be a quite different figure (as is indicated by social expenditure to improve environmental quality, which may well be substantially greater than the damage costs it is designed to eliminate). The value of certain environmental components are, indeed, almost impossible to assess directly in money terms, either for the benefits they confer or the damage their disturbance might cause: this is so for the biogeochemical cycles described in Chapter 1, for stratospheric ozone, for the decomposers breaking down dead plant and animal remains, or for the photosynthetic activities of green plants not taken as a farm or forest crop by man.

Such resources and processes in the natural world, which have not been created by society and appear to the non-scientist to be independent of man, tend to be taken for granted by economists as a 'free good'. The dispersion capacity of the environment used to be thought of in a comparable way. With mounting recognition that these natural systems are finite and capable of perturbation by man has come a tendency to treat them instead as social overhead capital (Uzawa, 1974), and recognise that this capital provides services used in productive processes and directly influences the level of economic welfare. Uzawa points out that the production and management of this capital is determined collectively by a society (this is where action to prevent possible perturbation of environmental systems by pollution comes in) while its services are provided either free or very cheaply to individuals and its maintenance is the responsibility of Government.

One way of valuing such capital is by the cost of maintaining it. Another would be to assess the cost of providing the services if this had to be done by the human community (for example, if we had to regenerate atmospheric oxygen chemically rather than via green plants). A third way is to assess values, especially for components like wildlife, historic monuments or beautiful landscapes, whose services to man are non-material, from social attitudes, recognising (as Figure 52 showed) that these vary with standards of living. Attempts to do this have not always been widely accepted (as with the valuations put by the third London airport enquiry on a twelfth-century church) and by their very nature are prone to argument, because there is bound to be a range of social values: the debate itself helps to determine the value appropriate to a particular time and community.

In principle, the economic approach can be used to compare options for alternative uses of the environment, including pollution-generating uses, and to determine the balance that represents the best use of natural resources. This approach demands evaluation, based on survey of environmental conditions, of the potential productivity of the environment for agriculture, forestry, raw materials, urban and industrial systems, wildlife, natural beauty and amenity. It is fairly easy to describe the present state of an area in terms of such parameters, and indices have been drawn up (e.g. Goldsmith, 1975) to bring together components like biological diversity and representation of rare species which may influence the value put on the area by those concerned with wildlife conservation (for a major national appraisal of this kind, see Ratcliffe, 1977). Helliwell (1969) has attempted to do the same for the features determining landscape quality, such as ancient trees: ancient monuments can similarly be brought into a landscape evaluation system. It is not impossible to predict how these values will change with various development options although this is not often done in land use planning. The difficulties arise when deciding what economic values are to apply to 'imponderables' like wildlife, history or natural beauty. In principle, however, this could be a way of guiding development so as to make the best use of the resources in the environment.

8.2.3 *Benefits*

A pollution-generating activity is not undertaken without reason, and that reason is the benefit it is expected to bring to the community. Commonly this benefit is expressed in terms of useful goods manufactured: sometimes it may actually be an environmental benefit (as when a sewage works replaces disseminated pollution from untreated human body wastes by more concentrated but more hygienic discharges). This benefit is the third term to be evaluated in examining the net cost of pollution.

8.2 *Evaluation of damage and benefits*

Some people might argue that the net damage of a polluting activity should be determined using a simple equation:

$$\text{Damage (net)} = \text{Damage (gross)} - \text{benefit}$$

If the benefits outweigh the gross damage, the activity is clearly a valuable one: if the converse is true, its social justification needs examination, for the community would appear to be making a loss on the deal.

Like damage, such benefits are difficult to compute. They include direct and obvious elements like the profit to industry on the products sold, or the gain to a country from the products exported and the employment provided. But they also include important elements like the stimulus a worthwhile job gives to individuals. In any event, the argument may be side-stepped here, for it is commonly agreed that while there may be little justification for an activity that imposes a net cost on the community because the harm it does (or the cost of controlling it) outweighs the benefits it brings, even beneficial activities should be controlled so as to balance damage costs and control costs in the fashion shown in Figure 51. This insistence is also regarded as equitable because it commonly happens that the damage caused by pollution may fall on different people from those enjoying the the benefits (thus air pollution from a factory may severely damage a farmer who buys only a trivial part of the products made, or water pollution from a brewery may destroy the fishing of a teetotal angler downstream). The optimum for the community as a whole is arrived at by balancing the social costs of the pollution against the social costs of pollution control (which include not only the actual costs of abatement, but an element reflecting the fact that the costs of control are likely to increase the cost of the products manufactured so that some people are no longer able to afford them: they thus lose whatever benefit the products bring). It is generally assumed that this 'fall in net benefit' is small compared to the costs of control equipment but there could be exceptional cases (as when the factory was making products designed to be available especially to poor people) where special measures might be needed to offset this cost, for example through subsidy. This would obviously apply if the products were important for public health.

8.2.4 *Control costs*

It is easier to cost actual expenditure on pollution control. A number of figures have been collected by Beckerman (1973) and by OECD (1972). In 1972/3 the United States figure was set by Kneese (1970) at $11 billion to $19 billion against a GNP of $1100 billion. Western developed nations

Table 28. *Estimated total pollution control expenditures in the United States (in billions of 1972 dollars)*

Pollutant/medium	1971			1981			Cumulative – 1972–81		
	O & M[a]	Capital costs[b]	Total annual costs[c]	O & M[a]	Capital costs[c]	Total annual costs[c]	O & M[a]	Capital investment	Total annual costs[c]
Air pollution									
Public	0.2	<0.05	0.2	1.0	0.2	1.2	7.1	1.4	8.4
Private									
Mobile[d]	1.1	<0.05	1.2	6.2	4.3	10.5	39.1	27.1	58.8
Stationary	0.4	0.3	0.7	4.2	1.5	5.7	27.0	11.4	38.4
Total	1.7	0.3	2.1	11.4	6.0	17.4	73.2	39.9	105.6
Water pollution									
Public									
Federal	0.2	NA[i]	NA	0.2	NA	NA	2.8	1.2	NA
State and local	1.2	3.8	5.0	2.6	7.0	9.6	20.0	47.2	76.9
Private									
Manufacturing	0.4	0.3	0.7	2.2	1.5	3.7	15.8	12.3	27.5
Utilities	0.2	0.1	0.3	1.6	0.9	2.5	10.9	6.8	16.5
Feedlots	0	0	0	<0.05	<0.05	<0.05	<0.05	0.2	0.2
Construction sediment[e]	<0.05	NA	NA	<0.05	<0.05	<0.05	<0.05	0.3	0.2
Total	2.0	4.2	6.0	6.6	9.4	15.8	49.5	68.0	121.3
Noise									
Commercial jet aircraft	NA	NA	NA	NA	NA	NA	NA	(0.4–1.6)	NA
Radiation									
Nuclear powerplants[f]	NA	NA	NA	<0.05	0.2	0.2	<0.05	1.2	1.0
Solid waste									
Public	1.0	0.2	1.2	1.7	0.4	2.1	13.8	2.6	16.3
Private	2.0	<0.05	2.0	3.1	0.1	3.2	25.2	0.3	25.5
Total	3.0	0.2	3.2	4.8	0.5	5.3	39.0	2.9	41.8
Land reclamation[g]									
Surface mining	NA	0	NA	0.8	0	0.8	4.5	0	4.5
Grand total[h]	6.7	4.7	11.3	23.6	16.1	39.5	165.2	112.0	274.2

From CEQ (1973).

[a] Operating and maintenance costs.
[b] Interest and depreciation.
[c] O & M plus capital costs.
[d] Excludes heavy-duty vehicles.

[f] Radiation figures include incremental costs only. The total costs of radiation control are inseparable from other costs of building and operating a nuclear powerplant.
[g] Land reclamation costs are assumed to be current expenditures.
[h] Does not include noise control.

generally appear to spend about 1 to 2% of GNP on pollution control. Figures for the United States are given in Table 28. In 1971–80 it is estimated that $106.5 billion was spent there on air pollution control, $87.3 billion on water pollution control, $86.1 billion on solid wastes disposal and under $3 billion on noise abatement. $61.0 billion was spent on automobile emission control as against about $45 billion on controlling pollution from stationary sources. This contrasts with the British pattern and reflects genuine divergence in national judgement of the relative hazard of oxidants and vehicle emissions on the one hand and smoke, sulphur dioxide and industrial pollutants on the other. Most of Europe concentrated on the latter. British figures are incomplete but Beckerman computed that in about 1970 there was an expenditure of about £100 million a year on sewage control (based on Jeger, 1970): there was probably a comparable industrial expenditure to prevent water pollution. Analysis of the 105th report of the Chief Alkali Inspector suggested that another £100 million per annum was being spent on air pollution control, very largely on industrial emissions and control of domestic smoke and very little on cars: making reasonable allocation for solid waste disposal the total emerged at around £500 million a year which was just over 1% of GNP. There can be argument over the details of such estimates, or the adequacy of the actions taken – but it would seem to have been established that pollution control costs have so far imposed only a small marginal burden on the national economy of industrial states.

8.2.5 *Financing controls*

When a factory (or a sewage works, for that matter) operates with only partial pollution control, emissions enter the environment, do damage and impose costs. This disseminated cost is spoken of as 'externalised': it is borne by people outside the pollution-generating activity and not concerned directly in it. This is commonly regarded as inequitable and it is now generally accepted not only that there must be controls to prevent unacceptable damage but that the costs of such control must be borne by the agency or industry generating the pollution: that is 'the polluter shall pay'. The result is that the costs of products bear an appropriate share of the expenditure needed to ensure that reasonable standards of pollution control were observed in their manufacture.

In practice, this internalisation of costs is normally applied whatever the standard adopted by the national or regional authority concerned, and whether or not it balances control and damage costs. The polluter pays principle stands apart from the economic arguments over the levels of control needed. Because damage costs are not, at present, easily assessed, and because value judgements in many communities are not strictly based

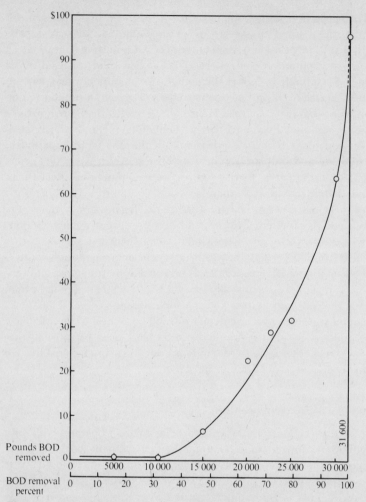

Figure 53. Incremental costs of reducing biological oxygen demand (BOD) content of lime flume and condenser water wastes from 2700-ton-per-day beet sugar plant. From CEQ (1971).

on economic or scientific assessments, standards are often set by the responsible control authority without a deliberate attempt at balancing damage and control costs. Yet the industries concerned are expected to carry the expenditure needed to meet those standards. The polluter pays principle is now accepted by all the member states of OECD, the Organisation for Economic Co-operation and Development (see Chapter 9).

As time goes by and knowledge grows, it may be that the standards of control set will move towards whatever the balance between damage and control costs indicates. Equally, damage costs may tend to be inflated by the increasing values an affluent community sets on environmental quality, while control costs may decline through new technology. It is desirable that this latter trend be sustained, for as Chapter 10 points out, as human populations and standards of living mount, the absorptive capacity of the environment will become increasingly used up, and if there is to be continued industrial growth it must be accompanied by increasing emission control – and the higher the standard demanded for the cleanliness of emissions, the more the control cost is likely to be (Figure 53). Predicting and costing future options may, for this reason, lead to somewhat higher investment in new pollution control plant when a factory is built, to avoid expensive 'retro-fitting' some decades later. Reduction in control costs, with gains in control efficiency, is an essential aim for human ingenuity if environmental protection and economic development are to continue to advance side by side as they have in many nations over past decades.

These arguments about matching control costs and damage costs, and ensuring that the costs of control are borne by the polluter, are also independent from the argument over *how* this goal is attained. In most countries it is done by regulation: that is by deciding the level of pollutant emission that is acceptable under particular conditions and then imposing a standard, a requirement to adopt a particular emission process or some other appropriate directive. It has been argued that it would be more efficient to levy a charge instead of impose a regulation, making the polluter pay for any emission above the specified limit: this would provide a strong incentive to adopt an efficient control system (but leave the polluter flexibility in choosing just how to do it), could more readily be adjusted to balance control and damage costs, and could also provide compensation for damage done (Royal Commission, 1972a; Rothenberg and Heggie, 1974). Some studies (e.g. Kneese, 1974) suggest that for water pollution a system of charging for effluents could achieve a specified standard at about one-half of the social cost of a system based on imposed effluent standards. These are matters of continuing debate but generally speaking charging systems are not in use as an instrument of pollution control today.

8.3 *Mechanisms for the control of pollution*

The objective of regulating man's use of the environment so that the best possible balance is struck, natural resources are conserved, biological productivity and ecological systems are sustained, and the damage done by pollution kept to the optimum level by economic controls is easy to state and widely accepted internationally. The actual ways of achieving such an objective vary widely.

Broadly there are two ways in which pollution is controlled in most countries today:

1. Through a land use planning or development control process in which the distribution of sources of pollution is adjusted so as to be compatible with other priority land uses, and so that pollution from new development is constrained from the outset;
2. Through controls, operated by various official agencies or voluntarily within industries, limiting existing sources of pollution and ensuring that new sources comply with conditions imposed when they are built.

These approaches reinforce one another. The precise organisations involved naturally vary from country to country: in the rest of this chapter particular stress is placed on the British system as an example, but parallels can be found in many other states.

8.4 Planning and assessment of the environmental impact of new development

8.4.1 The concept of environmental impact assessment

'Environmental impact' is a relatively young term. It has been introduced to describe changes in the environment which are caused by a particular human activity – such as the development of a new town, a highway or factory. In the United States, recent legislation (the National Environmental Policy Act of 1970) requires that the environmental impact of new developments undertaken by federal agencies or affecting federal lands must be assessed through production of an elaborate 'Environmental Impact Statement' (EIS). Guidelines for these have been promulgated by the Council on Environmental Quality (CEQ, 1971b; 1973). Even new administrative policies which could affect the environment are subject to this procedure.

The CEQ guidelines specify that the statement shall include:

1. A description of the proposed development;
2. A statement of its purposes;
3. A description of the environment affected;
4. The relationship of the proposed action to land use plans, policies and controls for the affected area;
5. The probable impact of the proposed action on the environment;
6. Alternatives to the proposal;
7. Statement of any probable adverse environmental effects which cannot be avoided (such as water or air pollution);

8. The relationship between local short-term uses of the environment and the maintenance and enhancement of long-term productivity;
9. Any irreversible and irretrievable commitments of resources that would be involved;
10. An indication of the other interests and considerations of federal policy considered likely to offset the adverse environmental effects of the proposed action.

Considering this list, it is not surprising that the resulting evaluations are sometimes voluminous: that on the pipeline to transport oil from the Alaskan North Slope to Valdez on the Pacific coast of the state took 22 volumes. Sometimes they have caused delays, especially where action groups have challenged the adequacy of an EIS in the courts (it is a practical impossibility, in any reasonable time and at reasonable cost, to describe every facet of the environment and predict with certainty the outcome of a development). Whatever the defects, the American policy has stimulated much thought about how to set out in an orderly manner the likely consequences of new development, and hence predict unwelcome effects and modify the development so as to avoid them.

Other countries have, naturally, adopted other systems appropriate to their administrative methods and legal and constitutional traditions. In the United Kingdom, Ministers (particularly the Secretaries of State for the Environment, for Scotland and for Wales) with their officials, formulate overall strategic goals and may themselves decide whether to allow proposed developments that are very large or pose particularly intractable or controversial questions. Within the UK system, local authorities carry major responsibilities both for broad strategic evaluations (structure plans) at county level and detailed examinations of individual applications for development (development control) at district level. British planning law is concerned with the control of changes in certain types of land use (especially the extension of urban or industrial development into the countryside, or switches from residential to office or industrial use of an area in a town) and does not control changes from one kind of agriculture or forestry to another: indeed these rural industries are almost wholly outside the planning system.

8.4.2 *The approach to environmental impact assessment*

In this chapter the term 'environmental impact assessment' is used to describe the broad process of examination of the effects of various changes in land use, especially where they may bring changes in pollution, rather than in the specific sense defined by United States laws. Pollution is only one topic to be covered in environmental impact assessment, because pollution is usually only one consequence of industrial development. The pollutants a new factory emits need to be evaluated within the wider context of environmental change due to that works, through its effects on land, communications, movement of people, demand for recreation and local economic patterns. Thus the environmental impact of pollution is commonly evaluated as a component of a longer 'check list' of effects. In a sense, a within-site survey of all the variables associated with the development is conducted, pollutants being among those variables. Sometimes, however, the impact of a single pollutant everywhere is examined in a kind of between-site survey. In such a case, overall assessments at national level are arrived at by an additive process once the field of variation in impact has been determined.

The likely impact of a development that emits pollutants depends critically on the amount emitted, the dispersion properties of the substances, the environmental characteristics of the receiving medium and the abatement technology that is possible, and any assessment has to cover all these aspects. It demands, as a result, a survey of the environment prior to development and an analysis of it as a functional system. Next comes the superimposition of the expected pattern of development – changes in the shape of the terrain, changes in the pattern of woods, fields, and urban zones, and changes in the levels of potentially polluting chemical substances that will be injected into the environment. The model constructed in this way can then be adjusted to test the impact of alternative development proposals including alternative pollutant emission points and concentrations – to provide a 'best fit' that causes minimal disruption to the features of the environment which it is sought to retain. Ideally, environmental impact assessment and this process of adjustment should not result just in a once-and-for-all statement gathering dust on a bookshelf, but in a dialogue between developer, environmental scientist and control agency extending up to the time when development is complete. This is the process followed for the Sullom Voe oil terminal in Shetland (SVEAG 1976) and generally in UK development control (Clark *et al.*, 1977).

The way environmental impact assessment is done in a particular country naturally reflects that country's approach to environmental planning. In Canada the Department of the Environment has surveyed large tracts of undeveloped land so that the social goals applied to each area can be decided well ahead of any impact. This 'pre-development audit' approach is obviously an ideal in parts of the world where there is much land in a natural state and time and money to expend on these kinds of extensive survey. It would be sensible in many developing countries if resources could be found. In Britain there is little unaltered land, but the process of structure planning, undertaken by county authorities, does demand the compilation of a broad statement about the preferred uses of every area and this too is being based on surveys of the existing situation, and evaluations of potential use. It is important to recognise that this survey is a two-step process. First, there is a need for accurate description of the environment today. This is provided by topographic mapping, geological survey, soil surveys, vegetation surveys, records of land use and agricultural and forest production, and surveys of associated fauna. Surveys of man's impact ranging from mapping the extent and nature of urbanisation and pollution to surveys of historic monuments are also part of the process. Environmental scientists would expect these surveys to characterise the natural resources of the area under study in some depth, and for this they should be objective — that is, they should sample and describe the situation in language that does not presuppose value judgements about what is to be done with the results (Holdgate and Woodman, 1975). Because environmental systems are dynamic, due attention must be paid to the time scale, and to the variations in climate, chemical factors and the cycles of animal and plant populations, and the construction of a simulation model of the main components is valuable.

The second step in the structure plan process is the development of policy objectives for each area. This draws upon the scientifically neutral surveys and models which define the areas expected to show particular changes in response to alternative developments, but leads on to value judgement about whether the change is welcome or unwelcome and to statements about the preferred environmental 'options' and their costs and time perspectives.

Much environmental impact assessment in developed countries like the United Kingdom takes place not in the predictive, anticipatory context of a structure plan or the Canadian surveys of potential land resource management but at the development control stage, when a specific new proposal is made. Such assessment is inevitably responsive rather than forward-looking: until proposals are made it is impossible to make any detailed predictions about what the consequences will be, because no two projects or sites are identical. At present there are several studies in Britain of how to improve the appraisal of the environmental effects of new developments within the context of existing planning law without needlessly lengthening the process of examination and approval. Several are reviewed in a recent symposium on environmental evaluation (PATRAC, 1976).

One approach (Catlow and Thirlwall, 1977; Holdgate, 1976) might lead at the structure plan level to the zonation of land into categories: 'A, highly valued and/or highly vulnerable'; 'B, valued and/or vulnerable'; 'C, not particularly valued or vulnerable'. More rigorous environmental impact assessments and development control procedures could then be concentrated on Zone A land – and its notification on a map to potential developers could serve as a warning that the case for their proposals will have to be more thoroughly justified there than on Zone C sites. The development and publication of the surveys would act, as existing planning maps do, to steer development to where it will do least harm.

8.4.3 *Pollution avoidance through development control*

The second way of improving matters is to help in codifying the questions to be asked of a particular development. One can start by classifying the developments (rather like land areas) into categories of potential impact. Scale and nature both come into the equation. A large factory (or town), or one releasing much effluent, is clearly more bothersome than a one-man affair. Works using processes scheduled under the Alkali etc. Works Regulation Acts or the Offensive Trades Acts clearly need more thorough attention than less troublesome types of industry. For various categories, needing varying kinds and degrees of scrutiny, a check-list can be produced. A manual proposing such procedures has been published by the Project Analysis for Development Control Unit at Aberdeen University (Clark *et al.*, 1977). Such a manual can contain keys in some ways like those used by biologists, but instead of reading 'white flowers, five petals . . . go on to question 8' it would say 'Process registered under the Alkali etc. Works Act? . . . consult HM Alkali and Clean Air Inspectorate and state agreed "best practicable means"' or 'Process emitting oily residues in water discharged to sea . . . consult Department of Energy and state acceptable emission standard and technology'. The aim is to ensure that before the development comes forward enough thought has been given to what it might do, proper agencies consulted, and the magnitude of its impact appraised (the United States guidelines also indicate the agencies to consult (CEQ, 1973). Such a checklist, by ensuring that awkward questions do not arise late in the process, when resources have been committed, might even speed up the granting of permissions. Figure 54 summarises the main components of a review leading to decisions about the type of environmental impact analysis that might be needed. Figure 55 illustrates the steps which, in theory, might be taken in assessing a particular development under this system.

The development control activities of planning authorities in Britain are supported and extended by those of certain other statutory bodies. All water abstraction and discharges to water courses demand consent from a Water Authority (when the Prevention of Pollution Act, 1974, becomes fully operational they will also control discharges to estuaries and the sea). The Water Authorities are concerned to ensure that sewerage systems are adequate to cope with the effluent from large new urban developments. Thus any plans for development are looked over by the Authorities who

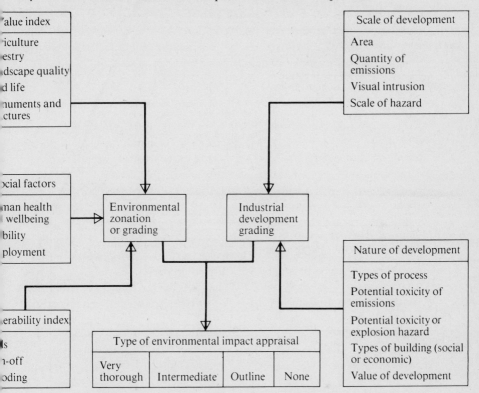

Figure 54. Components of environmental impact appraisal.

are enabled to seek information about the emissions expected and may impose conditions which must be met before they give their consent. The standard-setting process thus precedes development and can be used to ensure that the environment is protected from the outset. Similarly, all new works employing registered processes under the Alkali and other Works Regulation Acts must have the consent of HM Alkali and Clean Air Inspectorates. Before planning permission is given it is common for the Alkali Inspectorate to be consulted; if their verdict is that the 'best practicable means' available for a particular type of plant would not permit a satisfactory environmental condition under the proposed conditions of development, planning permission can be refused. Possible hazards to public health would naturally be considered by local authorities while the Health and Safety Executive would be consulted over possible dangers if there should be an accident in a proposed new factory (Committee on Major Hazards, 1976). At public enquiries into large industrial developments it is common for the impact of pollutant emissions to be debated and this has been a feature of a number of recent enquiries. In all these ways,

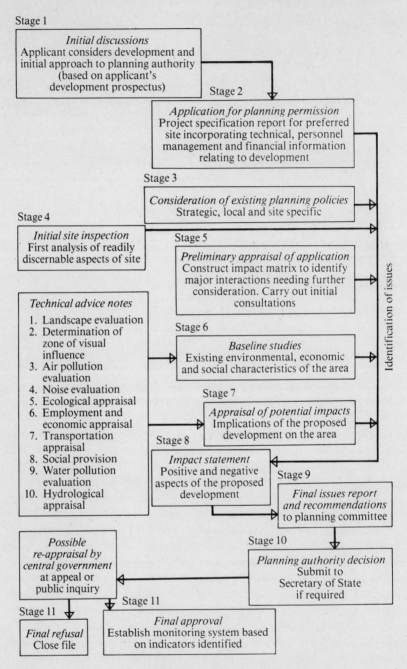

Stage 1

Initial discussions
Applicant considers development and initial approach to planning authority (based on applicant's development prospectus)

Stage 2

Application for planning permission
Project specification report for preferred site incorporating technical, personnel management and financial information relating to development

Stage 3

Consideration of existing planning policies
Strategic, local and site specific

Stage 4

Initial site inspection
First analysis of readily discernable aspects of site

Stage 5

Preliminary appraisal of application
Construct impact matrix to identify major interactions needing further consideration. Carry out initial consultations

Technical advice notes

1. Landscape evaluation
2. Determination of zone of visual influence
3. Air pollution evaluation
4. Noise evaluation
5. Ecological appraisal
6. Employment and economic appraisal
7. Transportation appraisal
8. Social provision
9. Water pollution evaluation
10. Hydrological appraisal

Stage 6

Baseline studies
Existing environmental, economic and social characteristics of the area

Stage 7

Appraisal of potential impacts
Implications of the proposed development on the area

Stage 8

Impact statement
Positive and negative aspects of the proposed development

Stage 9

Final issues report and recommendations to planning committee

Identification of issues

Possible re-appraisal by central government at appeal or public inquiry

Stage 10

Planning authority decision
Submit to Secretary of State if required

Stage 11

Final refusal
Close file

Stage 11

Final approval
Establish monitoring system based on indicators identified

Figure 5. An approach to project appraisal under existing development control procedure. From Clark (1975).

compatibility between new and existing development can be sought, and control levels determined at the outset.

It is obviously helpful if specific guidance can be given about the environmental quality to be sustained in an area proposed for urban or industrial development, and means provided for calculation of the degree of change a particular level of pollutant emission may cause. There have been several reports doing this for noise. In Sweden, noise standards have been proposed for various indoor and outdoor situations (Table 29) and tables published to indicate the recommended distances between housing developments and roads of various types, carrying a wide range of traffic densities (Table 30: National Swedish Board of Urban Planning; National Swedish Institute for Building Research, 1970). If modelling of the dispersion of air pollutants, and the concentrations likely to be attained under various conditions, can be improved, similar guidance may be possible in future on the relative location of housing and industry (cf. Reay, 1976, 1978). However, the variability of the environment is such that in most cases it is likely to continue to be necessary to adapt such general guidelines to the local circumstances of individual developments.

There is one particular limitation to planning as a pollution avoidance method. It is that the prediction of trends and of future levels is extremely difficult. It is even more difficult to forecast the trend in public attitudes. For example, when Heathrow Airport was opened, few people could have forecast the growth of air traffic and the advent of very large and noisy aircraft such as the first generation of commercial jet airliners. It is equally difficult to forecast the future impact of noise from a road system unless these is information about traffic growth, the impact on the usage of that particular road through the development of new townships, plans for other roads and for changes in available fuels that may affect the pattern of transport and the trend towards quieter or noisier engines. The planner sets a pattern for a century or more in the structure of towns and highways he advises, but changes in technology are more rapid, and old towns today provide a context for activities that would not be tolerated in new communities. For example, a high proportion of British works using dusty forms of lead are situated in densely populated urban centres (DOE, 1974b) because this was an acceptable activity in such places half a century ago. This is not true today and yet the pace of urban renewal is not so rapid that these old works will be phased out in the immediate future. To close them prematurely places a burden of compensation on the local authority. The standards of human sophistication have recently risen faster than the pace of urban renewal. So long as this continues, planning is never likely to be the sole effective means of pollution prevention. It must be reinforced by regulation of the sources already sanctioned and operational.

Table 29. *Recommended noise exposure limits for various situations*

	Effective level (dB A)	
Activity/premises	Day (0600–1800)	Night (1800–0600)
Sweden[a]		
Indoors		
Dwellings		
Living rooms, bedrooms	35	25
Other spaces	40	
Work premises	40	
Schools, etc.	35	
Hospitals, etc.	35	25
Outdoors		
Recreational areas	55	
	Time unspecified	
West Germany[b]		
Living rooms, internal	35	
Outdoors, outside living rooms	50–55	
France[b]		
Living rooms (proposed)	32–47	
Other rooms (proposed	42–47	
United Kingdom[b]		
Suburban		
indoor	42	
outdoor (implied)	62	
Urban centre		
indoor	47	
outdoor (implied)	67	
United States[b]		
Day rooms (indoor)		
(EPA)	42	
(HUD)	30–35	
Other spaces (indoor)	42	

[a] From National Swedish Board of Planning.
[b] From Langdon (1976).
Note: indoor values assume closed windows. A maximum outdoor value of 55 dB (A) will generally ensure an acceptable indoor condition of 35 dB (A). The figures do not refer to maximum sound peaks associated with the passage of noisy vehicles.

Table 30. *Distances from housing development to road centre corresponding to exposure levels recommended in Table 29*

Type of road and speed (km/hour)	Building height (storeys)	Number of vehicles per mean annual 24 hour period				
		2500	5000	10000	20000	40000
National distributor						
110	1	80	120	190	290	300
	3	120	180	300		
	6	150	250	300		
Regional distributor						
90	1	60	90	130	220	300
	3	80	140	220	300	
	6	100	180	300		
District distributor						
70	1	60	60	70	100	
	3	60	60	100	160	
	6	60	70	120	220	
Local distributor						
50	1	15	30	50		
	3	15	40	70		
	6	15	45	80		

From National Swedish Board of Urban Planning.
Note: all distances in metres.
These figures assumed sound quality is attained solely by use of wide buffer zones between housing and roads: in practice other methods (especially screening) are commonly adopted and save land. Advice on screening is also given in the report from which these figures come.

8.5 *Regulation of pollution*

8.5.1 *Central strategy and co-ordination*

The organisations for pollution control in a series of developed countries (Belgium, Canada, France, the Federal Republic of Germany, Italy, Japan, the Netherlands, Sweden, the United Kingdom and the USSR) have been compared by the Council on Environmental Quality (CEQ, 1973) whose own publications provide information about the system in the United States. In 1973, five of these eleven nations had a ministry with comprehensive pollution control responsibilities, and two more (Japan and the United States) had an agency with this role. One additional state, Belgium, had a Steering Committee chaired by the Prime Minister to co-ordinate five ministerial working groups covering aspects of environmental protection. In the Soviet Union the Hydrometeorological Service was reported as having overall monitoring and control functions for air, soil and water pollution: various ministries covered sectors of the environment and the State Committee for Science and Technology reviewed the whole environment field.

Variable though the picture was, it appeared evident that the majority of these countries had recognised the need for central government action: (a) to review the state of the environment as a whole; (b) to consider national environmental objectives (including standards, in many cases); (c) to co-ordinate the work of subsidiary agencies, often at local level.

Sometimes there was a dichotomy between the broad review group, setting strategic goals, and the enforcement agency supervising action: this is the situation in the United States where the Council on Environmental Quality advises the President on the state of the national and world environment and suggests actions, while the Environment Protection Agency has especial responsibility for seeing that the actions required by Federal law are taken effectively.

In Britain responsibility for central strategy rests with the Secretary of State for the Environment who co-ordinates national action as a whole and is himself responsible for executive action to control most types of pollution in England (in Scotland and Wales the respective Secretaries of State direct this executive action). The Central Unit on Environmental Pollution in DOE acts as a focal point, in support of this co-ordinating role.

Within DOE, the executive responsibilities of the Secretary of State are supervised by executive Divisions concerned with clean air, water, the disposal of solid wastes and the impact of noise, and these Divisions, together with CUEP, report to the Deputy Secretary (Environment Protection) as the senior British official in this field (DOE, 1976e). The Chief Planner in DOE has responsibilities for overall planning strategy. Outside DOE in England the Department of Trade retains responsibility for the control of oil pollution at sea and for aircraft noise; the Department of Transport for pollution from road vehicles, railways and craft using inland waterways; the Ministry of Agriculture Fisheries and Food for the control of agricultural chemicals like pesticides, for enforcing standards for food purity, and for the protection of fisheries; the Department of Health and Social Security provides medical advice and guides environmental health services; the Department of Employment, through the Health and Safety Executive, watches over conditions in the working environment and hazards from new installations, and the Department of Energy is responsible for the energy-generating industries. In Scotland and Wales, most of the functions which fall to DOE in England are the responsibility of the respective Secretaries of State (DOE, 1976e; McKnight, Marstrand and Sinclair, 1974).

In many countries the central strategic authority is advised by independent expert or representative groups. The Council on Environmental Quality in the United States is essentially an authoritative senior advisory body to the President, publishing annual reports transmitted to Congress by the President. In Britain the Royal Commission on Environmental Pollution is a similar independent group advising Government but also publishing reports which bring major issues directly before the public. In Britain there are also more specialised advisory councils for noise and for clean air, established before the Royal Commission and covering aspects of pollution prevention strategy in greater detail.

8.5.2 *Decentralisation to local government or specialist agencies*

In most countries there is a dual decentralisation of the executive action to control pollution – to both state or local administrations and to national specialist agencies. In countries with a federal constitution, like the United States, the Federal Republic of Germany, Canada or the USSR, State, Land or Republic authorities commonly have their own environmental policy and enforcement agencies. In Britain, the pattern is complicated and illustrates the diversity of agencies that may come to exist when environmental protection develops piece by piece in response to demand over the years (see DOE, 1976e for the most recent and comprehensive account).

Three kinds of body are concerned with the control of *air pollution* in Britain:

(a) A national agency, HM Alkali and Clean Inspectorate, is responsible in England and Wales for establishing 'best practicable means' for the control of emissions to air from over 2000 factories using over sixty industrial processes which produce particularly noxious or offensive emissions, or are especially difficult to control (Table 31). (In Scotland there is a separate but comparable Inspectorate.) Another national agency, the Factory Inspectorate, linked with the Alkali Inspectorate under the Health and Safety Executive, enforces standards in the working environment.

(b) Local authorities are responsible, under the Clean Air Acts, for control over emissions from other industrial processes and for controlling smoke emissions from domestic premises in areas which they define.

(c) The Department of Transport imposes regulations on vehicle emissions from both diesel and petrol engines (harmonising these as appropriate with EEC standards), and officials of the Department carry out annually more than 200000 spot checks on levels of smoke emission from vehicles.

Table 31. *Works and processes registered under the Alkali etc. Works Act*

(a) Number of works registered

	1971	1972
Alkali works	5	4
Scheduled and other works	1870	2166
Total	1875	2170

(b) Number of separate processes under inspection

Process	Number 1971	1972	Process	Number 1971	1972
			Brought forward	1114	1078
Alkali			Bromine	57	53
Saltcake	5	4	Hydrofluoric acid	14	15
Copper (wet process)	—	—	Lead	65	65
Cement	48	45	Fluorine	23	26
Smelting	19	16	Acid sludge	17	15
Sulphuric acid	4	3	Iron and steel	181	155
Sulphuric acid (class II)	41	46	Copper	51	49
Chemical manure	33	33	Aluminium	121	87
Gas liquor	36	25	Electricity	170	169
Nitric acid	111	117	Producer gas	17	10
Sulphate and muriate of ammonia	39	38	Gas and coke	146	132
Chlorine	123	119	Ceramic	253	220
Muriatic acid			Lime	30	31
(a) Other than alkali	150	150	Sulphate reduction	1	1
(b) Tinplate flux	4	3	Caustic soda	4	4
(c) Salt	2	2	Chemical incineration	19	25
Sulphide	125	128	Uranium	7	7
Alkali waste	—	—	Beryllium	11	11
Venetian red	—	—	Selenium	15	15
Lead deposit	1	1	Phosphorus	8	8
Arsenic	19	20	Ammonia	40	43
Nitrate and chloride of iron	3	3	Hydrogen cyanide	24	23
			Acetylene	3	2
Bisulphide of carbon	38	35	Amines	68	69
Sulphocyanide	—	—	Calcium carbide	1	1
Picric acid	1	1	Aldehyde	15	14
Paraffin oil	—	—	Anhydride	10	12
Bisulphite	108	108	Chromium	6	5
Tar	66	60	Magnesium	8	7
Zinc	26	22	Cadmium	23	22
Benzene	80	72	Manganese	4	4
Pyridine	32	27	Metal recovery	58	57
			Petroleum	60	58
			Acrylates	3	8
			Di-isocyanates	18	34
			Mineral	338	739
Carried forward	1114	1078	Total	3003	3274

From Chief Alkali Inspector (1972).

Noise pollution in Britain is partly the concern of local authorities, who are empowered to deal with noise nuisance under the Noise Abatement Act, 1960, and can designate Noise Abatement Zones and control construction sites etc. under the Control of Pollution Act 1974. The Department of Trade, however, retains responsibility for aircraft noise and under the Civil Aviation Act, 1971, may designate any UK airfield for purposes of controlling noise or requiring the payment of grants to help local residents install sound proofing. Vehicle noise is a responsibility of the Department of Transport: from April 1970 new vehicles have been required to meet set noise limits both at the manufacturing stage and whilst in use, while roads can be closed to, or special routes established for, noisy heavy vehicles. Standards for vehicle noise are now being harmonised throughout the European Communities. Finally, noise in the place of work (still the greatest hazard) is the concern of the Secretary of State for Employment's Industrial Health Advisory Committee and is likely to receive attention from the Health and Safety Executive in future.

Water pollution control is provided for under yet another system – a series of regional agencies, the water authorities, established under the Water Act, 1973, and responsible for the development, management, use and control of rivers and other aquifers within groups of river systems. These authorities set consent conditions for all discharges to inland waters, and under the Control of Pollution Act, 1974, their powers are being extended to estuaries and inshore seas. The consent conditions, as Chapter 6 explained, are emission standards adjusted according to the capacity, condition and use of the receiving waters. Generally, the authorities endeavour not to permit the quality of the rivers under their care to deteriorate: this means prevention of deoxygenation even in a dry summer with low river flow, and the consequent support of an ecosystem including coarse fish, and the maintenance of water safe for human recreational use. Today, the authorities are especially concerned to safeguard water for drinking, and now that an increasing amount is re-used, and there is concern over rising levels of 'trace' contaminants and 'micropollutants', this can demand very high effluent standards and the exclusion, through planning controls, of some kinds of discharge from certain watercourses.

In parallel with this statutory control system, there is one class of fresh-water pollutant, detergents, whose progressive regulation illustrates a distinctive British pollution control mechanism – a voluntary agreement between industry and government. An expert Standing Technical Committee on Synthetic Detergents (1960–77) has reviewed the environmental effects of these materials and advised on the phasing out of the 'hard' types causing foaming of rivers: these changes have been adopted voluntarily by the industry. A similar voluntary scheme operates in the *land pollution* field, where the Advisory Committee on Pesticides and other Toxic Chemicals has drawn upon information from industrial trials (and specified tests to be used in screening), suggested conditions to be applied for the use of pesticides in agriculture, horticulture and domestically, and stimulated the phasing out of substances whose persistence or side-effects were considered undesirable (ACOP, 1967, 1969). This Advisory Committee has recently broadened its work to cover chemicals used in wood and textile preservation, in water systems, and in other non-agricultural circumstances reviewed in a recent DOE Report (1974c).

The other main problem of land pollution is the disposal of toxic wastes. This is a local authority function under the Deposit of Poisonous Wastes Act, 1972, and Control of Pollution Act, 1974 and guidance on what is required, with codes of practice, has been issued by DOE in various circulars. Local waste disposal authorities are required to prepare waste disposal plans, and ensure that satisfactory arrangements are made in their areas, including the licensing of suitable sites. It is an offence to deposit poisonous, noxious or polluting waste anywhere on land in circumstances where it can cause danger to persons or animals or pollute water supplies.

This is yet another example of a situation in which the broad strategy for pollution control has been determined at national level by legislation, on the advice of a national working party of experts, but where the actual enforcement of the regulation has been decentralised to local authorities. To a degree this contrasts with the very special arrangements that have been made for the control of pollution by *radioactive materials*. Because of the need for a high standard of expertise in the handling of these materials stringent controls have been laid down and operated by central government agencies. Within nuclear power stations and fuel processing plants, the handling of radioactive material is directly controlled by the Department of Energy. Controls on the disposal of radioactive waste on land are maintained by the Department of the Environment, and by the Ministry of Agriculture, Fisheries and Food when their disposal at sea is permitted. Under the Oslo and London Conventions only certain categories of radioactive waste contained in special containers and deposited in deep water may be dumped in the ocean and the most intractable materials are stored in carefully shielded and segregated sites on land: in the future these may increasingly be located underground. These questions have recently been debated by the Royal Commission on Environmental Pollution (1976) and in its response (Cmnd 6820, 1977) the Government announced that responsibility for environmental aspects of nuclear-waste disposal, with the attendant research, was being conferred upon the Secretary of State for the Environment.

Marine pollution is the latest sector to come under comprehensive control in Britain. Until a few years ago the main control over new discharges came through the planning system. However, when the

Control of Pollution Act, 1974, is fully operational, water authorities will impose consent conditions on all emissions to the seas and estuaries in their regions just as they do for rivers. Outside inshore waters, the dumping of wastes is controlled by the Dumping at Sea Act, 1974, operated by the Ministry of Agriculture, Fisheries and Food (this allowed the United Kingdom to ratify the Oslo and London Conventions on dumping, described in Chapter 9). Oil pollution coming ashore, in contrast, remains a local government responsibility and emergency arrangements have been drawn up for the whole coast: central government provides grants to help in the clean-up operations and the Department of Trade, responsible for oil at sea, has ships equipped with dispersants on stand-by to deal with slicks that threaten the coasts or shallow seas valued for their fisheries or other resources. The Department of Trade also watches over British activities in response to the international conventions dealing with oil pollution, under the Prevention of Oil Pollution Act, 1971 and the Merchant Shipping (oil pollution) Act, 1971. Oily discharges from refineries are controlled by several means, including consent conditions issued by the Department of Energy under the Petroleum and Submarine Pipelines Act, 1975 (DOE, 1976b).

8.5.3 The overall pattern

Figure 56 sets out a kind of 'standard system' based on the experience of many states, for administering pollution control. The common features are:

(a) A minister responsible for overall strategy;
(b) Independent, authoritative advice on that strategy (with specialist sub-groups dealing with sectors or problems needing detailed attention);
(c) Some National Inspectorates or Control Agencies dealing with issues that require handling on such a basis (radioactivity is so treated in most countries: in federal states a federal agency may act as a 'watchdog' over state actions or police federal government and agency actions);
(d) Delegation to state or local authorities for general pollution control actions not requiring nationwide uniformity of approach or particular specialist skills;
(e) Linkage of pollution control in fresh waters to catchment management and water supply as a whole;
(f) Linkage between pollution control and land use planning;
(g) Organisation of monitoring so that it 'feeds back' to the control agency and overall strategic review groups.

The proportions of the elements can obviously be varied, and the

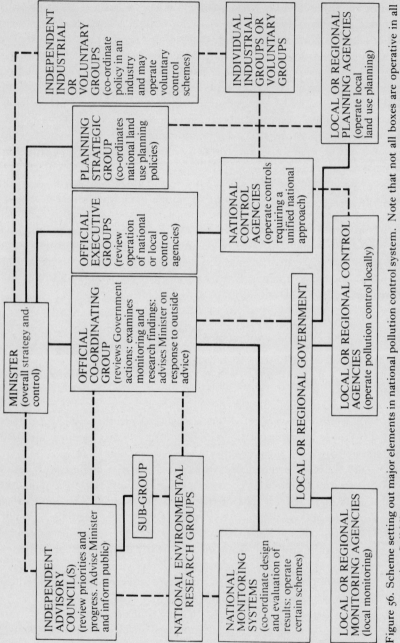

Figure 56. Scheme setting out major elements in national pollution control system. Note that not all boxes are operative in all countries. Solid lines represent formal reporting links and dotted lines major directions of information flow.

systems are still evolving. No single 'cook-book' recipe can be proposed: the most important requirement is to take account of national traditions and administrative patterns and adapt to these, and to be able to adjust to changing circumstances. Various 'feedback' capacities, especially through monitoring, research and risk assessment need to be built into the structure.

9. Pollution as an international problem

Pollutants originate from sources within or under the jurisdiction of individual states, but their pathways can be lengthy and involve the global systems of atmosphere and ocean. Pollution is evidently at least in part an international problem. In this chapter three questions are considered: (*a*) how numerous and extensive are these international problems of pollution? (*b*) what kinds of issue arise? (*c*) what kind of actions have been taken in response?

9.1 *The nature of the problem*

There are four categories of pollution problem which demand international action:

(*a*) pollution that transgresses national frontiers and affects the territory of adjacent states ('transfrontier pollution');
(*b*) pollution arising from mobile sources and products;
(*c*) pollution that affects the equilibrium of atmosphere and ocean;
(*d*) problems arising from differing national approaches to pollution.

9.1.1 *Transfrontier pollution*

Because many pollutants move with air and water currents, there is a considerable likelihood of their crossing a frontier. The scale of any consequent problem depends on the location of source and target, the properties of the polluting discharge (amount and toxicity of the material emitted) and the properties of the recipient environment (whether or not liable to favour dispersion because of terrain and current flow (Figure 57). In none of these variables is there any scientific difference between a problem of international pollution and one internal to a country. The same physical and chemical laws apply. The problems arise because actions in one country may cause damage in another (a fairly tangible issue that legal and administrative systems should be able to deal with), and because such pollution may use up environmental dispersion capacity and hamper the freedom of the recipient state to disperse its own pollution. The situation can be aggravated if there are differing economic and social priorities and legal systems in the countries of origin and receipt. These might include:

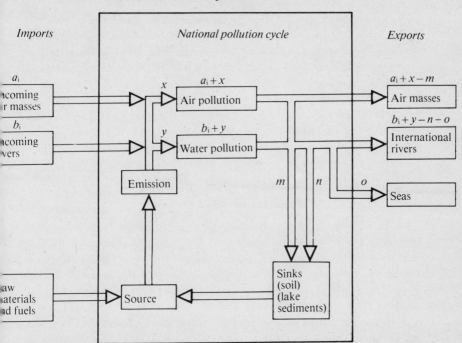

Figure 57. Imports, national cycle and exports of pollution. Note that critical policy decisions depend on whether imports and exports should balance.

(a) different estimates of the risks posed by the pollution, or the value of the targets damaged;

(b) different attitudes to the use of the absorptive capacity of the environment;

(c) a lack of agreement on how the absorptive capacity of shared air or water should be apportioned between states;

(d) different systems of controlling potentially polluting activities;

(e) different interpretations of the respective liabilities of government, industry and individuals for the damage caused by pollution, and differing legal systems involving (for example) varying standards for proof of evidence.

9.1.2 *Pollution from mobile sources and products*

The most familiar examples of this kind of international problem arise from: (a) pollution by emissions from motor vehicles; (b) pollution from aircraft; (c) pollution from ships (oil being the most familiar example); (d) pollution from manufactured goods like detergents, pharmaceuticals, pesticides, aerosols and many more.

As with transfrontier pollution, the problems do not differ scientifically from those posed by these same goods within their country of origin. They again arise both directly because damage may be caused and indirectly because environmental capacity is utilised. They are particularly acute when a vehicle or ship or aircraft or product designed to a particular standard passes into the air, land or water space of another country with either a more sensitive environment, or a higher level of existing pollution (so that incremental additions are less tolerable), or social policies which demand a higher standard of environmental quality. Again the problems are compounded by the possibility of differing responses in different countries with effects on mobility and trade.

9.1.3 Effects on the world environment

The third type of international pollution problem may arise if the emissions from sources under the jurisdiction of many nations collectively hazard a world resource, such as the balance of the atmosphere (with a threat of climatic change), or the ocean beyond the limits of national jurisdiction (with a threat to ecological stability and the fisheries of the high seas). The key questions are about how these resources (including the absorptive capacities of the environment) should be shared out, what quality goals should be set, and how, and what standards it is appropriate to impose. There is a fundamental question of entitlement: is any nation entitled to alter the climate to the detriment of other countries – and if it does, what liability is there for the damage caused? and on what time scale? If world climate today is threatened by rising carbon dioxide emissions beginning a century ago how far are nations today answerable for the deeds of previous generations who had no reason to suspect the impact of their actions?

9.1.4 Effects on trade and economic balance

The three preceding categories of international pollution problem are at least partly scientific: that is, they hinge upon the extent to which certain effects are taking place in the environment, and on which nations are responsible for the cause. Like pollution problems within a country, survey, research, monitoring and assessment are needed before policy can be determined: the difference is that these policies need to be extended to the international sphere. But the fourth type of international pollution problem stems from uneveness in just those social responses. Non-tariff barriers to trade may be created by, for example: (*a*) differing standards for the design of products e.g. detergent biodegradability or aerosol propellant composition; (*b*) differing standards for permissible levels of contaminants in foods; (*c*) differing standards for the design of mobile sources of pollution e.g. vehicle emission controls, tanker ballast systems or aircraft noise and pollutant discharge.

9.2 The scale of the problem

9.2.1 National and regional 'hot spots'

As the preceding chapters have shown, most pollutant effects are localised. Whether it is the warming of the air over London – a literal 'hot spot' – or the checking of crop growth around Los Angeles or of lichen floras on English tree trunks, the zones of demonstrable damage generally extend over only a few tens of miles. But many 'hot spots' overlap frontiers, which rarely coincide with natural barriers to pollution or regions of easy dispersion, like great mountain ranges or wide seas. Mobile pollutant sources can also aggravate 'hot spot' problems, for example if cars designed to emit relatively polluting exhausts are permitted to enter an area already more or less saturated with pollutant: that is why Los Angeles standards have been adopted as the general rule in the USA. Generally such problems are not hard to recognise, and many can be resolved bilaterally, or between small groups of countries.

Sometimes, however, the problem broadens to become regional. Much has already been said in Chapters 4 and 5 of the drift of sulphur compounds over Europe (and almost certainly over other industrialised regions). An international research programme under OECD auspices is examining the facts of the situation (DOE, 1976c) but the questions of damage estimation, national rights to unsullied air, entitlement to the use of the dispersion capacity of the environment, and how to assess costs and benefits on an international scale are all clearly subjects for future debate.

Similar regional problems are evident in great international river systems. The North American Great Lakes are well known to have deteriorated in water quality and biological richness over the past fifty years (although they are emphatically not 'dead'), and Canada and the United States have recently taken action to resolve the problem (Beeton, 1971; Smith, 1971; CEQ, 1972, 1973). The Rhine and the Danube drain many states and receive effluent from them all, but the state at the seaward limit naturally has the accumulated effluent to contend with. It is possible to measure how much pollution crosses the frontier (e.g. Wolff, 1978) but action cannot be taken there. The Germans cannot refuse to receive the Rhine from Switzerland, or the Belgians and Dutch to take it over when the Germans have finished with it. Nor is it practicable to set up a major pollution removal plant at the frontier. The contamination is only likely to be abated economically by treatment at the myriad sources within each country sharing the drainage system of the river. Regional standards and regional cleanliness thus depend on individual controls, taken in individual states and enforced there.

The Baltic, the Irish Sea and the North Sea are all examples of shallow, moderately land-locked waters, where pollutants discharged by several countries accumulate, and while there is no proof of damage to fisheries, hazard to health or severe ecological upset in any of them, the threshold of such damage is certainly not so far off as to make complacency tolerable (Sweden, 1972; ICES, 1969; ICES, 1970; DOE, 1971; NERC, 1975; Sibthorp, 1975). There does seem evidence that the deaths of 12000 seabirds in the Irish Sea in 1969 may have been aggravated by the loading of polychlorinated biphenyls in their bodies (Chapter 5). Mercury levels likewise are highest in fish from inshore waters, such as the coasts of the Baltic: in Britain recent surveys (MAFF, 1971, 1973) showed that compared with average levels in oceanic fish of around 0.05–0.1 parts per million, North Sea fish averaged around 0.3 ppm and fish from the Thames Estuary and eastern Irish Sea, offshore from known sources of contamination from industry, averaged over 0.5 ppm. Like the Great Lakes or the Rhine, these seas are thus examples of regional shared resources, unequally polluted by several states.

Although the dilution capacity of the ocean is enormous, various persistent pollutants such as DDT, polychlorinated biphenyls and heavy metals have been discharged in sufficient amounts to cause international concern (FAO, 1971). Accumulation may be slow but because amelioration would take a comparable time scale, a large safety factor is clearly prudent. Oil is another widespread marine pollutant whose impact on fisheries and marine ecosystems has not at present caused widespread damage (Chapter 5; FAO, 1971; GESAMP, 1977), but which is an unwelcome disamenity, and it is understandable that nations that neither produce, transport nor consume it in quantity resent its arrival on their shores.

Chapters 3 and 4 described how global and regional climate might be modified by carbon dioxide, released from the burning of fossil fuels and from forest clearance, and also by changes in cloud cover and the concentration of particulate dust in the stratosphere. If these kinds of effect, originating in the actions of many nations (not all of them developed countries), were indeed to alter the boundaries of agricultural production and the margins of deserts, major questions of international liability would inevitably arise. Similar questions are posed by the possible modification of stratospheric ozone concentrations as a consequence of high-flying aircraft, aerosol propellants, and oxides of nitrogen from many sources.

Radioactive contamination could clearly be another international problem of no mean order, for there is no true lower threshold of effect, and all the evidence confirms that any rise in the contamination of atmosphere or ocean with long-lived radionuclides is a threat. This problem was recognised several decades ago with the formation of the United Nations Scientific Committee on the Effects of Atomic Radiation (UNSCEAR) and the International Commission for Radiological Protection. So far, threats have come mainly from the testing of nuclear weapons, but peaceful uses of nuclear energy are certain to increase and some could cause problems of contamination by accident (Royal Commission, 1976). There will certainly be continued international monitoring of all such potential sources of global pollution.

Pollution is clearly, therefore, an international problem at both regional and global level. So far, regional problems are more acute and more frequent, but the global issues pose in many ways the most intractable problems and cannot be ignored.

9.3 Questions to be answered

9.3.1 Kinds of question

If these problems are to be solved (as they must be, and are being) it is clearly necessary to answer several different and difficult questions:

(a) How should levels of contamination in areas of shared environment be measured?

(b) How should the resulting data be analysed so as to ensure international agreement over its conclusions?

(c) How should risk estimation approaches be harmonised internationally so as to promote agreement on the scale of hazard (as distinct from the scale of effect)?

(d) How should the absorptive or dispersion capacity of the environment be used, and how should this capacity be apportioned between states?

(e) How should liabilities of states for transfrontier pollution be assessed, appropriate international compensation levied, and action to abate the polluting activity promoted?

The need to respond to such questions internationally was stressed in the Declaration of the United Nations Conference on the Human Environment, held at Stockholm in 1972 and the precursor to the United Nations Environment Programme. It said

The protection and improvement of the human environment is a major issue which affects the well being of peoples and economic development throughout the world: it is the urgent desire of the peoples of the whole world and the duty of all Governments... the discharge of toxic substances or of other substances and the release of heat, in such quantities or concentrations as to exceed the capacity of the environment to render them harmless, must be halted in order to ensure that serious or irreversible damage is not inflicted upon ecosystems... States should take all possible steps to prevent pollution of the seas by substances that are liable to create hazards to human health, to harm living resources and marine life, to damage amenities or to interfere with other legitimate uses of the sea...States have in accordance with the Charter of the United Nations and the principles of international law the sovereign right to exploit their own resources pursuant to their own environmental policies, and the responsibility to ensure that activities within their jurisdiction or control do not cause damage to the environment of other states or of areas beyond the limits of national jurisdiction.

9.3.2 *Scientific questions: research, monitoring and evaluation*

Chapter 7 has described how international monitoring schemes are being developed under the auspices of organisations like the United Nations Environment Programme and the World Meteorological Organisation, with scientific advice from SCOPE (the Scientific Committee on Problems of the Environment) and other professional groups. The evaluation of the resulting data forms an integral part of such monitoring schemes, and UNEP is supporting basic studies of the methods of monitoring and assessment, some of them at the Monitoring and Assessment Research Centre (MARC) (SCOPE, 1977).

These global schemes are augmented by a large number of regional activities, some specifically related to international pollution problems. The monitoring of water quality in the Great Lakes, the Baltic, the north-eastern Atlantic and the Mediterranean has been or is being designed as an outcome of the bilateral Great Lakes Water Quality Agreement (CEQ, 1972), an agreement between states bordering the Baltic (Baltic Convention, 1973), the Oslo Convention (Command 4984, 1972) (which is drawing upon the expertise of the International Council for the Exploration of the Sea), and the Convention negotiated at a conference in Barcelona sponsored by UNEP (1976). The Organisation for Economic Co-operation and Development (OECD) is, as mentioned above, the co-ordinating body for research on the dispersion of sulphur oxides over Europe, and the outcome of these studies will certainly be discussed widely in European international groupings. The state of the Rhine is being monitored under an international agreement that established a Rhine Commission and the European Economic Community is taking steps to link the monitoring networks of member states, as well as supporting a substantial scientific research programme.

These studies are only part of a major international scientific effort. In the 1960s and 1970s many international organisations began to debate pollution problems. To name but a few:

(a) OECD set up an Environment Committee which began to meet in 1970;

(*b*) The Economic Commission for Europe has set up a group of senior advisers to Governments on environmental problems;

(*c*) NATO established a Committee on the Challenges of Modern Society with many environmental activities, including studies on low-pollution vehicles and advanced sewage treatment, and has held seminars and advanced study institutes on environmental problems;

(*d*) The Council of Europe has been concerned with the side-effects of pesticides and other toxic chemicals since the middle 1960s;

(*e*) UNESCO has included pollution research in its Man and the Biosphere programme;

(*f*) FAO and WHO have been concerned with the effects of pollutants on agriculture and human health;

(*g*) The International Council of Scientific Unions co-ordinates the scientific community in a number of programmes concerned with environmental issues (SCOPE, 1977).

By and large, it is not difficult to organise international programmes of scientific study and monitoring, partly because the methods are often available, and partly because there are so many existing international organisations, both at inter-governmental and professional level, in which such issues can be discussed.

9.3.3 *Questions relating to the use of the environment as a sink*

The duty on states not to engage in acts which might jeopardise the balance of global environmental systems was stated clearly at Stockholm, for example in the report of the Sessional Committee on Pollution:

It is recommended that Governments be mindful of activities in which there is an appreciable risk of effects on climate and...carefully evaluate the likelihood and magnitude of climatic effects and disseminate their findings to the maximum extent feasible before embarking upon such activities...consult fully other interested states when activities carrying a risk of such effects are being contemplated or implemented.

This resolution implies establishment of a basic goal – not to disturb global systems. Short of this point, there are of course many activities which discharge wastes safely to the environment. As Chapter 6 pointed out there are two basic philosophies about pollution. One is that any additional loading of the environment with potentially hazardous substances is bad – in economic terms a disbenefit – and that the only correct goal is the same purity of air and water as prevailed before man became industrialised and so great a burner of fuel and discharger of sewage (see, for example, Davis, 1971). In international terms, it might be held, as a development of this view, that each nation is entitled to receive 'pure' air and water from its neighbours, and is under an obligation to pass it on in similar condition. The other view is that it is just as legitimate to use the capacity of the environment to disperse wastes as to use any other natural resource, but that we must not misuse this resource: that is, we must set limits to the damage we will accept. This concept prevails in Britain and most developed countries, and even in the USA where the conservation movement has sought decrees that the National Environment Policy Act prohibits any development that causes any decline anywhere in the environmental quality, the courts have upheld the view that some deterioration in 'clean' areas is acceptable as part of a re-location and modernisation of industry if the overall trend is towards improvement. But there are considerable arguments over how much of the absorptive capacity of the environment it is legitimate to use, and its apportionment among users is an even thornier problem at the international level. Should the use of this resource, for example, be entirely internalised within a state: that is, should all discharges be taken up by internal sinks or should some export of pollution be allowed as a right comparable with the right to use the common resources of sunlight, air and water?

The first step might be to decide what emissions to the environment are tolerable. This demands understanding of pathways and effects, the process of risk estimation, and an agreed evaluation of the benefits to be gained from the polluting activity and the costs of its consequences. Even within a single country this is a matter of debate, for judgement of the value to be put on a target that is hazarded is not easy, and there will inevitably be divergences about the benefits of the activity generating the pollution. Internationally, all these steps become that much harder. This is especially so because the costs and benefits are rarely likely to be evenly balanced. A new industrial complex may bring great benefits to the country within which it lies, but none whatsoever to the neighbour state. Indeed it may bring disbenefits: its emissions may partly 'use up' the absorptive capacity of air or water within that neighbour state's territory, thereby imposing an increased control cost on its domestic industry, while there may be trade competition from its products. It is easy to see how feelings can run strongly.

To ease this problem it would be valuable were there some international agreement about air and water quality objectives (another way of stating agreement on the use of the absorptive capacity of the environment) for the shared environment. This might lead on to an agreement about how far the shared resource might be apportioned. There would, inevitably, be many elements to take into account in reaching such agreement, including traditional usage of the resource and the availability of alternatives. It is possible that one way of analysing the situation would be to establish a notional 'value' for the resource and an apportionment of both the contributions to that value made from each side of the frontier and of the relative benefits derived from its use: these might both be brought into a notional costing scheme which could lead on to a financial adjustment if the situation appeared inequitable.

This would, perhaps, be easiest to do at regional level, where only a few states were involved. There are no actual examples where this has been done, so the whole issue remains conjectural: perhaps the nearest illustration is the Great Lakes Quality Agreement between Canada and the United States (CEQ, 1972) where both countries have agreed on general and specific water quality objectives and on steps to reduce pollution. In this example, however, each nation is bearing its own costs, and there does not appear to have been an effort to allocate future use of the resource, although this could presumably be done by the Great Lakes Water Quality Board set up under the agreement.

A hypothetical example may illustrate the general point in the absence of a real case study. Ruritania and Harmonia share Lake Aquaria, each owning half the surface. The lake is not, however, of uniform depth, the Ruritanian half containing a deep trough while there are broad shallows on the Harmonian side. Ruritania, it can be calculated, owns 60% of the water by volume. Against this, 80% of the inflowing water, on which the lake depends for its continuing dilution capacity, comes in from the mountains of Harmonia, and is, moreover, scarcely polluted whereas the one significant Ruritanian river is already contaminated to the point where its trout fisheries are impaired. The other pollution of the lake arises in approximately equal proportions from twin towns on either side of the lake which shed their sewage into its waters.

At the conference table the parties might agree on a series of quality objectives. Knowing the quality of the unpolluted natural inflows and the time it takes for the lake volume to 'turn over' it is possible to calculate the safe dilution capacity for various substances. Assume that this dilution capacity is divided up into 1000 arbitrary units. At present, Ruritania uses 300 of these, 200 because of the pollution on its incoming river and 100 due to town sewage: Harmonia only uses 100 units. The Ruritanian claim might well be that their country is entitled, on a proportionate volume basis, to 600 units and uses 300, leaving them 300 while by coincidence Harmonia, entitled to 400 and using 100 also has 300 in reserve. Both can therefore embark on an equal industrialisation programme – a thoroughly equitable situation! On the other hand the Harmonian delegate might argue that his country supplied 80% of the water in the lake and should therefore enjoy 800 units of its dilution (after all, there is nothing to stop them polluting the river while within their territory prior to inflow, as Ruritania does, leaving no dilution capacity at all) while Ruritania should only have

200 units attributable to its inflow. On this basis, Ruritania is using 100 units too many already, should not undertake any development increasing the polluting load, and should pay Harmonia a rent for the 100 units of Harmonian capacity it is using. A formula to resolve the issue is clearly needed and it might be decided to balance inflow and volume and allocate according to the arithmetical mean, giving Ruritania 400 units and Harmonia 600. An arrangement under which one country could lease capacity the other did not need might also be discussed: a Joint Aquarian Commission to administer a monitoring scheme would be a foregone conclusion.

9.3.4 *Questions of souces and targets*

Pollution causing international problems may be produced by individual citizens, by public agencies, or by industries, and different solutions are likely in each case. Only governments (central or local) can regulate the mass of emissions from domestic chimneys, vehicles, or sewers. Transfrontier pollution caused by such sources cannot be traced back to any individual, so that if damage is caused, only the central or local government in the country of origin is likely to be answerable: should there be legal insistence on finding a person to whom blame can be attributed, the search is almost certain to fail. On the other hand, an individual can clearly be held responsible for the noise and emissions of his vehicle, once it crosses a frontier into another state, and he must expect, in default of some treaty granting exemption, to be required to meet the standards prevailing in the country he is visiting, or be penalised as its laws prescribe if he fails.

Works operated by government agencies or industries are easier to deal with, if only because they generate pollution in larger amounts at fewer points. They can clearly be held responsible for transfrontier pollution that can unequivocally be traced to them, although it may not be easy to disentangle the effects of a single source out of many. How, for example, should responsibility for transfrontier pollution by sulphur dioxide be apportioned, when it comes from so many domestic and industrial sources?

In such circumstances, a government may find it easier to seek redress from a neighbour government, as responsible for its citizens, than proceed against a particular industry, unless the case is clear-cut. Mobile sources and products are easier to deal with: a manufacturer can be required to meet product standards, or accept responsibility if his product is shown to cause damage. Industry, too, can abate pollution caused by products directly, by voluntary means, as well as be required to do so by statute, giving further flexibility. This has been done for certain classes of 'hard' detergent, and in Britain for polychlorinated biphenyls: international companies have the ability to operate similar controls throughout their area of operation and linkages between the companies in a whole major industry may resolve a problem in advance of government action. The oil industry provides examples of this, in promoting the 'load on top' system for tankers, to reduce marine pollution from oily ballast and tank washings, and in TOVALOP, the Tanker Owners Voluntary Agreement on Liability for Oil Pollution, which guarantees compensation for damage caused by oil as a kind of industrial mutual insurance scheme.

Clearly, in many cases, it will be virtually impossible to prove that a particular injury due to transfrontier pollution has been caused by a particular individual controlling a single source in the country of origin. Redress may consequently be hard to obtain through the courts, because there is nobody to sue. Hence the general trend in this field towards intergovernmental agreements on standards and controls that prevent the problems arising.

9.3.5 Standards

One way out of the problem of variable laws and consequent disputes might be the establishment of universal standards. If all nations agreed to observe the same standards for quality of air and water, it might seem that the basic problems would be solved – for such standards would automatically define how much pollutant could be passed on from country to country. They would also set limits to the extent to which pollution of atmosphere and ocean was regarded as tolerable on a global scale.

But there are several problems here. Uniform standards or goals of environmental quality would allow very variable controls on individual emissions (Chapter 6; Figure 48). In areas with no pre-existing pollution, quite dirty discharges would still, if adequately dispersed, not give rise to excessive concentrations. Conversely, in heavily industrialised areas, very stringent emission controls might be needed if the quality standards were not to be exceeded.

It can be argued that in scientific terms this would be sensible. It would also favour developing countries with great environmental capacity, and impose the heaviest burden of pollution control costs on developed states. But such flexibility would inevitably be opposed by those who prefer to see uniform emission standards for a particular process everywhere – either in order to impose equal costs on industrial competitors, or so as to maintain the quality of those areas of environment which are at present only slightly polluted.

This whole area of trade distortion and non-tariff barriers, which causes the most heated arguments in the international pollution field, applies most of all to products This is because products, like vehicles, can go anywhere. If consumer protection demands that nobody eats fish with more than 1 part per million of mercury in it, and that paint should never contain more than 0.05% lead – both sensible standards – this must apply to all fish or paint sold in a country. So must a standard requiring all synthetic detergents to be 90% biodegradable in the environment – again a sensible provision to prevent the foaming of rivers. But this implies that the manufacturing industry everywhere must comply or forfeit exports to the state setting the standard. The effects were seen in the USA when tuna fish were found to contain more mercury than the Department of Health, Education and Welfare considered safe. There was widespread prohibition of sale and import – but because exports were not banned, there was considerable dumping of American canned tuna on states without this standard. On the other side of the coin, economic protectionism and environmental protection can very easily get mixed up. In 1971 a bill was presented to Congress that would have excluded from the United States imports of foodstuffs from countries whose policies for use of organo-chlorine pesticides were less stringent than in the USA – and this was criticised as protectionist by nations whose trade it threatened.

It is understandable that one approach favoured by many governments is to seek agreement on the basic levels of protection needed by people everywhere. Such basic health standards – primary protection standards – have been set by the United Nations Scientific Committee on the Effects of Atomic Radiation and the World Health Organisation has advised, for example, that water should contain no more than 20 ppm of nitrate, and air should contain no more than 250 micrograms smoke and 500 micrograms sulphur dioxide per cubic metre (Chapter 6). It is likely that an increasing number of such international guidelines will be agreed: in addition there may be more stringent secondary standards or higher goals to which nations move as their wealth permits. These international goals would be achieved by national regulation of emissions, using whatever machinery is appropriate under the prevailing constitutional and administrative circumstances. Where there is need to harmonise practices like dumping of wastes at sea or release of oil from tankers in order to achieve agreed objectives, special conventions can be drawn up – or special commissions like that for the Rhine established. Product standards can be set with the agreed protection standards also in mind. Because these may interfere with trade, they need international discussion before adoption. The OECD has established, for this purpose, a procedure whereby a member government gives early warning of new measures it is contemplating that might have an adverse effect on the flow of goods. In the EEC, member states similarly notify the Commission of forthcoming measures, and if these seem likely to affect the Community as a whole a member government may be asked to wait while a co-ordinated directive for the whole Community is worked out. Such procedures provide a foundation on which to build.

Such an approach may be less readily pursued on the global scale. Although scientific study and international risk evaluation may bring the scientific community to agree on the broad scale of a potential problem (as the World Meteorological Organisation has recently done over the possible effects of carbon dioxide), it is another matter to translate this into action. Standards at global level are not easy to set (at least in a meaningful form). Taking carbon dioxide as an example: models suggest that a doubling of carbon dioxide might raise average world temperature by about 1.5 °C (Chapter 4). This might be considered the maximum change tolerable, and a standard of 600 ppm, for carbon dioxide in air, averaged globally might be decided on. The models suggest that it would be exceeded if fossil fuel consumption continued to rise at present rates, by around 2010 AD. It is at such a stage that the problems begin. If a brake is to be put on fossil fuel consumption, how should it be applied? Some might argue that the developed countries have the means to substitute other energy sources most readily, as well as being the main energy users and hence (since it takes about 50 years to being about major changes of this kind) that they should now embark on a programme to substitute nuclear, solar, tidal or other energy sources. But in what ratio? What form of nuclear power is to be favoured – and what standards should be imposed on it? Should limits be set to the permissible fossil fuel usage per head of population at prescribed future dates? At what stage in the process of development should new countries become subject to such standards? Should there be restrictions on forest clearance, also a source of carbon dioxide? These are truly intractable problems and it is not surprising that at the present time the emphasis remains on scientific study and assessment so that the risks are really defined before nations embark on the massive investment of effort and money that major changes in energy supply patterns would require. Moreover, there are both finite limits to exploitable fossil fuels and other incentives to conserving them because of their value as raw materials in the chemical industry, and many people believe that technological forces now operating are likely to bring about changes in the energy supply system 50 years from now without the need to impose an elaborate series of international standards and regulations in order to prevent pollution.

9.3.6 *Questions of harmonising enforcement policies*

Once an approach has been agreed, however this is done, questions may arise of how to enforce it, and of how the costs should fall. This is very largely a matter for internal decision within the countries concerned, but it may be argued that if different ways are adopted of financing the pollution control needed to avoid unacceptable damage, there may be economic imbalances that also spill over into trade. For this reason the OECD nations have recently agreed to adopt the 'polluter pays principle' (see Chapter 8), and this has been included as an element in the EEC Environment Programme. The adoption of this principle internationally means that distortions to trade are minimised: the OECD states, which include all the western developed nations with a market economy form a fairly self-contained trading system. The essence of the 'p.p.p.' is that it is a 'no-subsidy' system. This is not complete for there is recognition that exceptions may be needed, for example where an industry giving essential employment would be priced out of the market if it had to bear the costs of putting its effluent in order. Social factors have to be allowed to remain a matter for the discretion of individual governments. The broad principle however is of harmonisation of approach so that a reasonable standard of environmental protection is attained without distortion in product costs and aggravation of trade wars. There is no case for more harmonisation in this area than in those others where divergent government policy affects the costs of products. But the evidence is that overall there is a fairly even balance. OECD figures, for example, sugest that most western developed countries spend about 1–2% of their gross national product on environmental protection.

As Chapter 8 also stressed, marginal costs have to be balanced against marginal benefits. The cost of increased pollution control should balance the benefits it brings in reduced corrosion, dirt, or health hazards or enhanced crop production. As new technology allows more efficient abatement techniques, the standards of control can and should naturally rise. Conversely, as control costs rise – for example through raised energy costs – we may be forced to allow some decline in pollution standards. But here we come up against international value judgements again. Countries differ in their levels of development and in their costing of both control and benefit. If there is mass unemployment, squalid housing, no medical services and mass starvation, and a fertiliser plant will help relieve these miseries, the equation will balance at a point that guarantees a substantial output of product even at a cost of environmental damage which would be unacceptable in countries with a higher standard of living. Sauce for the goose is not necessarily sauce for the gander, and while the world remains so unequal a place economically and socially, it may well be logical that it remains unequal in national policies and priorities for the environment. We cannot, therefore, expect differences to disappear simply through better scientific assessment of damage, or agreements on how to apportion international resources, though both help to make issues clearer.

9.4 International action to control pollution

9.4.1 Laws and conventions

International scientific study, monitoring, agreement on standards and on broad principles for the use of the 'environmental sink' and a certain harmonisation of economic approaches help to reduce international pollution problems to more manageable proportions. But they do not solve all the issues. Ultimately, questions of law and enforcement must be answered.

Most developed countries have laws – some, like the European and North American States many laws – concerned with pollution. At international level there are three needs:

(a) harmonisation so that the differing laws of states do not actually aggravate differences in environmental quality in frontier zones or shared areas of environment;
(b) agreement on how to resolve disputes over infringements of agreed standards, or damage caused by transfrontier pollution;
(c) joint action to fill gaps in legal controls or to legislate jointly where a common resource needs protection.

The Traill Smelter case is the most commonly referred to when liability for transfrontier damage due to pollution is discussed. This factory in British Columbia was undoubtedly the cause of damage to vegetation and environmental quality in the United States. Legal argument hinged on how far liability extended. It was agreed that there was a right for compensation for established damage, but not for losses of environmental amenity which could not be costed. It would probably be this precedent that would be cited in any international legal dispute, for example over any demonstrable reduction of southern Scandinavian fresh-water productivity by sulphur dioxide from southern Europe – though here the ambiguity over sources might well complicate matters. In general there seems to be an emerging consensus that where an air or water mass moves from one country to another, the receiving country is not entitled (in default of a specific agreement on quality standards) to demand that this water or air be any cleaner than the standard demanded in its own territory. For example, if a state sets an air quality standard as 100 μg SO_2/m^3, an air current over its frontier polluted to 75 μg/m^3 must be accepted: it is legitimate for the neighbouring state to behave in this connection as if it were part of the receiving state. It would not be reasonable for the latter to impose on industry in its neighbour's territory higher standards than it imposed on its own. This is a practical way out of the difficulty over defining how much pollution may legitimately be exported – leaving the way open for specific negotiations if it is regarded as unsatisfactory.

Problems arise where the opposite situation prevails: where the discharging state has a lower standard and the incoming air is dirtier than the recipient state allows domestically. There may be legitimate reasons for the difference: the environment in the state of origin may generally have a greater dilution capacity and industry be less dense, so that there has been no need for a stringent standard, or there may be other socio-economic factors which lead the first state to accept a higher margin of disamenity or damage than the more affluent second can tolerate. Be that as it may, the recipient is likely to argue that the state of origin should raise its standards or pay the costs of the marginal damage in the recipient state accounted for by the difference in the two standards. If the state of origin is a developed country of comparable standing with the recipient it may seem that this case is fair: difficulties would arise if the recipient were a rich country and the pollution originated in a poor one. In this case there is a widespread feeling in the United Nations that it would be inequitable to tax a country already suffering from a lack of wealth in order to safeguard or compensate a rich country which has, because of its favoured circumstances, been able to set high environmental standards. This argument has yet to be resolved.

Conversely, when we turn to shipping, many developing countries feel that it is basically inequitable for them to be exposed to the risk of oil pollution or hazard from toxic cargoes carried past their shores by the fleets of the wealthy nations. At the Law of the Sea Conference and its various preliminary meetings such countries have argued for the right to establish 'pollution control zones' extending some 200 miles from the coast, and to set standards for shipping operating there to ensure that hazards to their coasts and coastal fisheries are minimised. There are precedents for such action. Canada, in 1970, legislated to give herself powers to control shipping over her very extensive, ice-beset continental shelf. Britain, in the Oil in Navigable Waters Act of 1971, confirmed her power under customary international law to intervene on the high seas where a casualty in an accident at sea appeared to pose a threat to her coasts. These are examples illustrating concern which has led to national action: the Law of the Sea Conference has however yet to secure world agreement on the major principles involved in balancing the interest of coastal and shipping states, developed and developing countries, and the interests of defence, commerce, fisheries and mineral exploitation in the seas of the world.

These examples illustrate a still fragmented situation over the har-monisation of national laws and the settlement of disputes. Alongside, numerous bilateral and international measures have been developed to deal with particular pollution issues.

For example, several international conventions on marine pollution have been signed and ratified. There has been an International Convention for the Prevention of Pollution of the Sea by Oil since 1954 (amended in 1962) and this is also affected by parts of the Convention on the High Seas signed at Geneva in 1958. The main Convention controls the discharge of oily mixtures from ships, especially tankers. The oil industry has now developed the 'load on top' system which cuts oil pollution in tank washing by some 80%. New conventions developed under IMCO, the Intergovernmental Maritime Consultative Organisation, deal with accidents to shipping, ship design, hazardous cargoes and ships' wastes.

The other international pollution problem at sea to have been of wide concern is the dumping of hazardous wastes like those identified by GESAMP in the 'black list' of materials whose discharge to the marine environment should be minimised (Chapter 2). In 1971 an Intergovernmental Working Group established by the Preparatory Committee for the Stockholm Conference met in London and then in Ottawa: further meetings followed at Reykjavik and at the Stockholm Conference itself before a meeting in London in the autumn of 1972 drew up and adopted the London Convention on the Dumping of Wastes and other Matter (HMSO, 1972). A similar Convention, restricted to north-west Europe, had been adopted at Oslo earlier in 1972 (Command 4984, 1972). Both conventions demand the minimising of dumping of the most noxious materials in the ocean, and care in the disposal of other, less hazardous substances. They establish a consultative framework for the participating states. They set up a monitoring system. Their scope was restricted to dumping at sea but in 1973/74 the north-western European states went further in adopting the Paris Convention which extends controls to the discharge of hazardous materials from the land – via rivers, pipelines and coastal outfalls. In the Baltic, the riparian states reached agreement on pollution control in a convention signed at Helsinki, while a further far-reaching convention on marine pollution, this time for the Mediterranean, was signed at Barcelona in 1976.

In parallel with such conventions for the sea, others have covered inland waters. A convention on the Rhine establishing a commission to set standards for management of that river has been operational for some time. The Council of Europe in 1974 drew up a more wide-ranging convention on the protection of international rivers from pollution. In 1972 a bilateral agreement was adopted by Canada and the United States to improve water quality in the Great Lakes. In the EEC the Environmental Programme includes demands for directives that will unify community regulations for control of priority pollutants. The evidence is therefore that where problems have been clearly enough defined and debated, international action is following.

There is good evidence that these actions work best on a regional scale. It was quicker and easier to negotiate the Oslo and Paris Conventions than the London Convention on dumping and, whereas all the north-west

European states have signed the local conventions and a number have ratified, several major shipping states have not signed or ratified the London Convention.

9.5 *Conclusions: the pattern of international action*

The evidence of this survey can be summarised as follows:

First, most countries agree that pollution is of concern for the damage it causes. Our aim is to confine that damage to reasonable levels. As standards of living have risen, the acceptable limits have tended to become more stringent.

Second, a survey at the present time suggests that most pollution problems concern local 'hot spots': some are more regional, affecting rivers, air masses or seas within or adjoining several states: only a few are potentially global. There is thus an especial need for bilateral or regional international action.

Third, these international issues arise not only because of damage which may be caused to an environmental resource shared by several states, but because the action taken to contain such damage could distort trade and establish non-tariff barriers. There is thus an economic as well as scientific strand in the argument.

Fourth, at international as well as national level the elements in the problem and our response to it involve discussion, to agree the outlines of the problem, scientific study to improve our understanding of it, legislative harmonisation in conventions and treaties to express common agreement, standards as specific elements in the agreement defining the goals or limits all agree to be reasonable or the procedures all agree to adopt, and monitoring to determine whether success is being attained. In all these fields economy of effort is important: the linkage of national efforts in a network is more effective than the supra-national establishment of some overlord body. As Figure 58 indicates, national and international action must reinforce one another, if only because much of the research and assessment and almost all the regulatory action must be national.

Fifth, there are auguries of success in these fields. The process begun at the Stockholm Conference is still incomplete. The developed and developing countries differ in their priorities. But we are definitely moving in the right direction. Where a pollutant has been shown by science to pose a real problem, its control will usually be found to be under international discussion, and commonly some action will be evident. It can be argued we are not going far enough and fast enough, and that self-interest (not all of it enlightened) prevails at international as well as national level. But there is enough movement for cautious optimism.

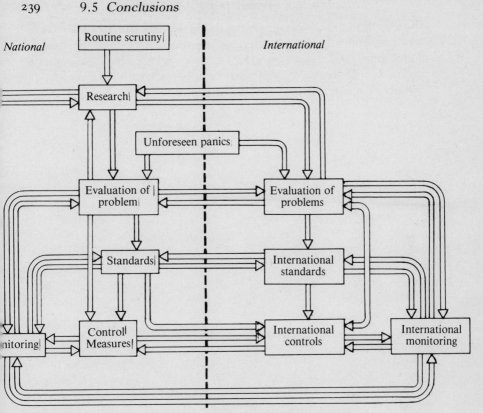

Figure 58. National and international responses to pollution problems.

Lastly, a cautionary note. We must not confuse pattern with process: the static with the dynamic. The world, like all ecological and social systems is continually changing. Pollutant levels are on both rising and falling trends. It is essential that control expands at least as fast as the problems develop. Our rate of response must always exceed the rate of change – preferably by a substantial safety margin. We cannot afford to drift from panic to panic and we have done this too much in the environmental field. It took a *Torry Canyon* to stimulate faster action to combat oil pollution at sea. It took 12000 deaths of seabirds in the Irish Sea to confirm the side-effects of polychlorinated biphenyls and two years later secure their voluntary withdrawal by a responsible manufacturer. It took Minamata to alert the world, through some 80 deaths in horrible circumstances, to the hazards of mercury. So far, what has happened is that an acute 'hot spot' situation has provoked strong national action which, through the network of communication and debate internationally, has been copied in other states and spilled over into joint action. In future, through the development of 'early warning' systems like the UNEP Global Environmental Monitoring System and the International Register of Potentially Toxic Chemicals (IRPTC), we may hope both nationally and internationally to detect threats before they materialise, and respond fast enough to spare people suffering and hazard.

10. The approach to the future

10.1 *Is pollution a threat?*

The concern expressed about pollution in recent years has many causes. But behind many of the more disturbing warnings of doom has lain a fear that if pollution continues to increase it could threaten the quality of human life, if not the future of man on earth. These are issues intimately bound up with the growth of human populations, and the demands that mounting numbers of people make on the natural resources of the planet.

10.1.1 *Human population growth and its limitation*

Various projections and speculations suggest that human numbers, at present doubling about every 30 years (Figure 59) will for some while continue to follow this exponential curve, which is paralleled by exponential increases in energy consumption and urbanisation (Ehrlich and Ehrlich, 1970; Ward and Dubos, 1972: National Academy of Sciences, 1971; Royal Society of Arts, 1973). Limitation, which must ultimately come in a finite world, depends on the interaction of many factors including social adjustment and birth control, shortages of natural resources and food, and pollution. The preliminary models of Meadows and others (Meadows *et al.*, 1972) suggested that human numbers might 'overshoot' and collapse within the next 150 years because natural resources became limiting: if this collapse was avoided by doubling the available resources then pollution might become limiting. Well short of such limitation, severe stresses would inevitably be imposed on the structure of societies and the quality of life.

Several authors have criticised the design of these models and disputed their conclusions (see, for example, Ashby, 1975 for a very readable comment). More 'optimistic' views of the population and pollution issue (e.g. Maddox, 1972; Holdgate, 1973) commonly make the assumptions (some factual and others more akin to pious hopes) that:

(*a*) Pollution at present is a negligible factor in human mortality, even in the worst polluted areas.

(*b*) Generally, there is a substantial safety margin.

(*c*) Population increase has not, in past experience, necessarily been

241

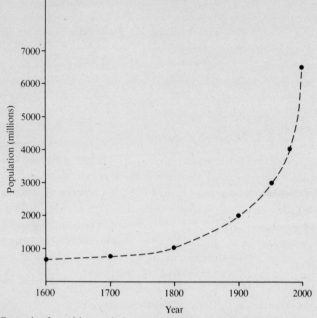

Year

Figure 59. Growth of world population (estimated for 1980 and 2000). After Ward and Dubos (1972).

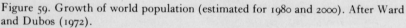

correlated with increased pollution emission or with increased mortality from this cause.

(*d*) This is because rising populations have either been agricultural, generating negligible amounts of toxic substances, or if industrial, able to impose pollution abatement technology that compensated for the increase in potential emissions.

(*e*) Population growth has, in the past, been slowed most effectively as material standards of living have risen, through the development of voluntary birth control: this slowing down comes some time after medical science and technology have reduced the death rate, especially in infants, thereby reassuring parents that even a small family will suffice to protect them in age.

(*f*) Human societies are not stupid, and exponential curves do not continue indefinitely in the real world: numbers of people have been limited in the past and a balance will be struck before catastrophe ensues on the global scale.

This 'optimistic' approach can be translated into models if birth and death rates are allowed to be modified by increases in wealth, which also allows pollution abatement or avoidance. In its simplest form, this thesis states that as long as rising numbers are accompanied by rising income and industrial investment, adequate expenditure can be devoted to pollution abatement: this does not consume so much wealth that it impairs the net benefits, and there is therefore also a rise in material standard of living which in turn provides for higher life expectancy and voluntary population limitation. There do appear to be historical examples of such a process in most western European countries during and since the Industrial Revolution. First death and then birth rates have declined in the same broad period as rising industrial output and per capita income: in this same period the urban environment has been improved both through better housing, drainage, water and food and through pollution control. But equally obviously there are still many countries where the initial 'trigger' to these processes, in the shape of industrial development, has not happened. In many developing countries rapid population growth is outstripping both agricultural and technological advance (Ehrlich and Ehrlich, 1970; National Academy of Sciences, 1971). Birth control campaigns have only brought a minor check to population growth in such countries, and there is an increasing recognition that these campaigns only work when they form part of a much more comprehensive process of development, removing the social and economic causes of over-population. Such matters lie outside the scope of this book, but one general conclusion does appear reasonable. It is this. Should the poorest countries fail to bring their populations into balance with the resources of land, food and wealth available to them, their numbers are likely to be limited in the end by famine, disease and conflict rather than by pollution.

This is so because agricultural and poor urban populations generate only relatively simple and natural forms of pollution – smoke, food residues and the body wastes of man and livestock. These are readily taken up by biogeochemical cycles, posing problems only when concentrations become very great – as when lakes and rivers become grossly contaminated with organic matter. Even then damage, for example to fisheries, is localised and it is the threat to health through the spread of disease because water supplies become fouled that is most serious. The environmental hazards of extreme poverty stem much more from the loss of food production because land is cleared badly, leading to the burning off of humus layers and the erosion of topsoil, or through the extension of settlement onto flood plains and coasts prone to natural disaster, than from what we in the industrialised world recognise as pollution. Even the addition of carbon dioxide to the atmosphere, recognised in Chapter 3 as the form of pollution most likely to cause world-wide climatic change, has come as much from the burning of fossil fuels to power the industries and heat the homes of the developed countries as from the cooking fires, huts and forest clearance of simpler communities.

10.1.2 *Pollution and mortality in man*

It is an evident fact that acute injury from pollution has killed people in many parts of the world. In 1952/3 the London smog caused some 4000 excess deaths. Even this dramatic incident had a negligible effect on national death rate, however: most of the victims were elderly people with respiratory weakness and no great expectation of life (Royal College of Physicians, 1970). The mercury poisoning in Japan, especially at Minamata, killed around 100 people and crippled several hundred. Extreme incidents like these have happily been rare, and deaths from such causes have certainly (as Chapter 6 indicated) been much less than from accidents on the roads and in the homes of developed countries, and from self-inflicted damage caused by smoking and drinking. Acute injury from pollution of the general environment has almost certainly also claimed far fewer lives than toxic chemicals to which people have been exposed at work, or toxins that have contaminated foodstuffs, as in the mercury poisoning incident in Iraq reported by Bakir and others (1973) or the many cases of fungal infestation of badly stored grain (Austwick, 1975).

Recent concern over the possible hazards to health resulting from pollution has focused far more on chronic than acute injury – probably because it is clear that chronic damage is harder to detect and to prevent. There are (as Chapter 5, and especially Section 5.2.4 illustrate) considerable margins of uncertainty. Some authors maintain that 60–90% of cancers are environmentally induced, that is, they are triggered by the biochemical action of substances in tobacco smoke, food, water and other drinks, cosmetics, drugs and pharmaceutical products or air, or by the direct impact of radiation (UNEP, 1977; Higginson, 1975). Industrial exposure to siliceous dusts in mining and quarrying, to asbestos and perhaps other fibres, and to a wide range of substances including vinyl chloride monomer, 2-naphthylamine, arsenic and tars, has been shown to shorten life in a proportion of those affected and cause much individual suffering. Section 5.2.4 and the report by Doll and others (1978) cite other cases where such exposure has apparently affected reproduction and caused abortion. Complex interactions may also occur in which pollutants increase susceptibility to other hazards: sulphur dioxide and radiation apparently enhance the susceptibility of smokers to lung cancer (Doll and Hill, 1974; Lambert and Reid, 1970). Human susceptibility to stress is affected by many factors including genetics, nutrition, age, personal history, climate and living environment. Some of these factors are liable to perturbation in a manner we are forced to treat as random, and there are thus major uncertainties in analysing the detailed impact of pollution on human populations and ecology – as there are, at this level of detail equally for other species and ecosystems. The mortality statistics may therefore conceal the true magnitude of the chronic effects of pollution, and only long-term research is likely to clarify the situation. However the epidemiological records surveyed by Doll (1978) do not provide any grounds for claiming any major increase in death or illness that can be attributed to pollution in Britain, at least over recent decades, and much for believing that in this period, despite rising population and industrial development, public health has steadily improved.

10.1.3 *The establishment of safety margins*

Even though we can be confident that acute injury to man will only rarely result from the pollution of the general environment, it does not mean that such risks can be ignored. It is natural that a safety margin be sought (Chapter 6), and there has been a considerable debate over its present magnitude. For smoke and sulphur dioxide, the evidence of the London smog (Figure 60) is that excess deaths became manifest when ambient levels attained 500 μg SO_2/m^3 in the presence of 250 μg smoke/m^3 (considerably more sulphur dioxide is tolerated in the absence of smoke) (Royal College of Physicians, 1970; Lawther, 1973). Such SO_2 concentrations may still be reached occasionally, for a few hours or even days, in urban areas, but averaged over the year concentrations are at half these levels or less. Rural areas exhibit one-fifth to one-tenth of the 'hazard' concentrations and remote areas of the world, probably around one-hundredth. Over the world as a whole there would appear to be a tenfold to hundredfold safety margin for these pollutants – though many people still live and work where margins are narrower.

For mercury, latest analyses (SCOPE, 1977) suggest that toxic effects become detectable when blood levels reach about 1 μg/g body weight. If a tenfold safety margin is allowed, giving a primary biological standard (Chapter 6) of 0.1 μg/g for a 'standard' man, this would be the equilibrium resulting from the dietary uptake of 50 μg methyl mercury/day. Surveys (MAFF, 1971, 1973) indicate that the average dietary intake in Britain, where there is a substantial consumption of fish, the foodstuff highest in methyl mercury, is not more than 10 μg/day and not all of this will be in the methyl form. Butler concludes that the safety standards currently in use in North America, Japan, Germany, France and Sweden are adequate to protect average populations and the only problems arise in groups with excessive consumption of contaminated fish. The standards are set at around 0.5 ppm of mercury in fish and while a few species (pike, and probably tuna) can reach this or higher levels naturally, oceanic populations generally contain only around one-tenth of this figure (MAFF, 1971, 1973).

Most people obtain the majority of their lead intake from food (in which

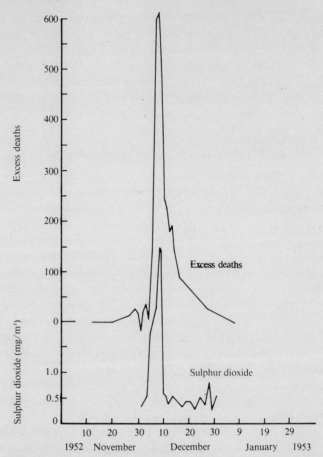

Figure 60. Daily deviations in deaths in Greater London (calculated as differences from the general trend), December 1952. From Lawther (1973a).

the metal is naturally present, as it is in soil and rock), and there is no evidence of a problem here. Water contaminated by the solution of lead plumbing (see Section 4.5.2) can however raise the body intake perceptibly, and this is probably a more significant problem than the much more publicised effect of airborne lead from motor vehicles and from industry. The latter, in a 'worst case' urban environment may contribute 0.3 mg/week, as against 1 mg/week from food and a further 1 mg/week from water if the latter is at the maximum standard recommended by the World Health Organisation of 0.1 mg/litre (DOE, 1977c). Even in these 'worst case' situations, symptoms of acute toxicity are unlikely, and the elimination of paints with high lead concentrations and of lead from pottery glazes and cooking utensils (DOE, 1974b) must over recent years have significantly reduced lead poisoning hazards.

The main pathways of potentially hazardous materials to man undoubtedly involve food and drink, with the air as a third and generally lesser
route. In Chapter 7 the kinds of monitoring programme adopted in order
to watch over the movement of materials of known hazard along these
pathways were reviewed. The other general conclusion is that the
environment in which some people work, especially in industries making
or using chemicals known to be hazardous but also in some other
apparently safe places like hospitals, may be a more dangerous one than
the outdoor or domestic worlds. Safety limits are set as a routine in many
industries (Chapter 6) where workers exposed to materials of known hazard
are under continuous medical surveillance. It is probably true that the
greatest opportunities for reducing environmentally-induced cancers and
other disorders still exist in the working environment and through the
better screening and control of materials that may enter or be added to
foods or are used deliberately by people on themselves or others, including
in medicine. Such safety standards can only be set substance by substance
and circumstance by circumstance.

10.1.4 *The capacity for pollution control*

Industrial development has undoubtedly increased emissions of toxic
substances and caused severe local damage, but because it has also been
accompanied by increasingly efficient technology for the control of
pollution, population increases associated with industrialisation have not
led to a universal parallel rise in environmental contamination. Some
pollutants (such as carbon dioxide) continue their upward trend. Others
have followed such a pattern initially, but then levelled off or begun
to fall (Chapter 4). Some very unpleasant areas of dereliction have undoubtedly been created, like the areas of devastated forest around smelters
in parts of Canada and the patches of soil contaminated by earlier smelters
in south Wales. But twentieth-century industry has generally been
characterised by enhanced output and enhanced pollution control. Most
forms of industry evidently *can* be kept clean: arguments are much more
over the level of cleanliness that is needed, and the costs it is in
consequence reasonable to incur. A safety margin is continually being
created by the development of new means for emission treatment.

It remains true that if the environment is of finite capacity but the
numbers of people and density of their industry increase, a 'crowding
factor' operates. At the most basic level this can have a direct effect on
birth and death rates, although man is an animal capable of sustaining
reproduction in astonishingly congested circumstances, and it may be
suggested that the more significant operation of the crowding factor is in
agriculture, where high densities of people per acre demand very high
productivity. In industry, the effect of density is to reduce the available

margin of dilution and dispersion capacity in the environment (Chapter 6 develops this point). If a river has a finite flow, as all have, and its quality is not to deteriorate, doubling the volume of discharges to it from bankside factories must demand a halving of the concentration of pollutant permitted in those discharges – or twice as stringent an emission standard for that effluent. Similarly, it is the great increase in motor vehicles in the cities that has necessitated the increasingly stringent control of their emissions.

The effect of this factor is to increase the cost of pollution control – and not in a simple way. It is usually much more costly and troublesome to take the last 10% of pollutant out of an emission than the increment from 80 to 90% (Figure 53). Moreover, as standards of living rise so people demand increased environmental quality. Similarly, enhanced scientific understanding of some of the less obvious chronic effects of substances used in, or released from, technology is undoubtedly likely to create demands for new controls. These factors operate to reduce the net rate of production of wealth from industry, as that industry expands. Theoretically, a state might be reached where the cost of human and environmental protection took all the net benefits. But most developed countries spend only a small part of their gross national product on pollution control (Chapter 8). There is a substantial margin, continually broadened by the ingenuity displayed in the development of more efficient pollution control technology at an economic cost.

10.1.5 *The threat from pollution: a tentative conclusion*

At the level of global generalisation used in the crude model of Meadows and others (Meadows *et al.*, 1972) it seems likely that the overall pollution burden in the world could certainly double, and maybe rise tenfold before acute human mortality rose significantly. But all the analysis in this book militates against global generalisations of this kind. Substances differ in their toxicities, pathways, persistences and interactions and need to be judged separately. Only a few are truly global in the scale of their dispersion and action. Most exert their effects over very small areas, acting on small populations exposed to unusually high concentrations, in 'hot spots'.

Tentatively, it is possible to suggest that the 'optimistic' approach outlined at the beginning of this chapter is more credible than the opposite judgement of the 'gloom and doom' school. It does seem true that industrial development has in many countries allowed a great increase in human quality of life and longevity, with, at the same time, a slowing in population increase and increasingly effective pollution control. It seems clear that pollution today is not a major brake on human numbers or a major cause of human misery. There are, of course, risks: the more potentially hazardous materials we use and the greater the scale of their production the more we must rely on our own skills at screening, containment and safe disposal and the less we can rely on the dilution and dispersion capacity of nature. In a finite world there must ultimately be limits to human numbers, and there are certainly limits to the concentrations we can allow the substances we make to attain in air, water, soil or food. We could certainly poison ourselves if we wanted to – just as we could bomb ourselves to destruction. But there seems no reason to believe that we must.

10.2 *Future trends in our approach to pollution*

There have been changes in both social and scientific attitudes to pollution over recent decades. To begin with, people were most concerned with the acute effects of toxic substances like the hydrochloric acid released from the early alkali industry, which led to the first Alkali Act, or more recently with the London smog, Minamata mercury poisoning, changes in the Great Lakes, lead in paint, water, food or air and photochemical oxidants in Los Angeles and other cities.

These effects are reasonably clear-cut. Their causes are readily identified. Damage can be measured and related quantitatively to measured exposures and doses. Control is feasible because sources can generally be identified, and since the effects are acute, there is unlikely to be a very long delay either between the manifestation of damage and a social response or between the imposition of controls and the improvement of the environmental situation.

There is no doubt that efforts to prevent acute damage from particular toxic materials must continue. As Chapter 2 indicated, many lists of substances with the potential to cause hazard have been drawn up and it is possible to recognise the properties that make a substance liable to become a troublesome pollutant. Screening of new substances that might be hazardous is becoming increasingly stringent. Information about such substances is being collected and stored in more sophisticated form, and the data banks are being linked to provide a world-wide network.

At the same time, new problems and attitudes are emerging. It is now generally accepted that the generation, dispersion and impact of pollutants must be seen in a wider ecological context. Pollutants, as chemical and physical factors, act on complex systems whose structure and operation are determined by many variables.

Because of this environmental complexity it is not easy to extrapolate the results of laboratory tests and experiments into the field. The pathways pollutants follow are complicated. Transformations into more (or less) hazardous substances can be very important. So can local accumulation by organisms.

The impact of a pollutant also depends on the condition of the target. Age, nutritional and reproductive state, genetic variation and the history of previous stress all modify susceptibility. So does the presence of other pollutants, for additive and synergistic effects are increasingly evident. It is becoming clear that the concept of a threshold below which a pollutant is contained by the homeostatic capacity of the target organism may need review. Apparent thresholds arise not because a pollutant has no biochemical effect at low doses, but because we have not yet developed sufficiently sensitive means of detection.

Attention is also focusing on the consequences of chronic exposures to low doses. Various statements have been made about the proportion of cancers with environmental causes. These are not easy to validate experimentally, especially if many years elapse between the exposure of the target and the appearance of the response. But epidemiological techniques, correlating the patterns of mortality and morbidity in man and other organisms with the pattern of exposure to pollutants, alongside patterns of variation in other environmental factors, can suggest hypotheses for detailed investigation (such methods established the relationship between smoking and respiratory disease, and between air pollution and the diversity of lichen floras).

In the future the kind of multivariate analysis used by the ecologist and the epidemiologist is likely to be employed to explore these complex circumstances of exposure of man and of other targets. In place of action to constrain or eliminate the causes of acute damage, we may see action through positive planning to sustain, or where necessary to create, combinations of environmental circumstances that are most likely to be healthy. Prevention is commonly held to be better (and cheaper) than cure, although the costs need not, in practice, be less.

The scale of the pathways and effects with which we are concerned is also broadening. Many pollution problems of today extend much wider than the acute hazards listed at the start of this section. Chlorofluorocarbons and carbon dioxide disperse globally in the atmosphere. Chlorofluorocarbons and DDE (the product to which DDT is converted in the environment) are only slowly degraded so that most of the material made by man remains in the biosphere for decades. Changes in climate and in the penetration of the atmosphere by radiation could be felt world-wide. Even short-lived pollutants, such as sulphur dioxide, are known to be capable of drifting for hundreds or thousands of kilometres.

It follows that pollution, which could be recognised, studied and remedied locally when we were largely concerned with acute effects, is now a matter of increasing international research and debate. This trend is almost certain to continue. There is bound to be regional discussion and concerted action to protect shared rivers, lakes and inshore seas and the regional pattern of drift of air pollution will also be of concern to the nations of the industrial zones. At global level, impact of pollutants on atmosphere, climate and oceanic systems is bound to be the subject of co-ordinated programmes of research and evaluation, as well as action where this can be taken.

The two dominant trends – toward the recognition of larger-scale and longer-period problems and toward an increasingly international scientific approach – are likely also to compel changes in social attitude. The change from relatively clear-cut acute effects, manifest in fairly unambiguous clinical symptoms and capable of exploration using well-defined laboratory toxicological experiments, to more general statistical correlations depending on multivariate analysis like that used in epidemiology and ecology is bound to increase uncertainty. Ecologists are used to the study of systems some of whose components behave in a manner best treated as random, and to the fact that many such systems do not have one natural equilibrium (and may indeed, lack any enduring balance). Assessments are inevitably stated in terms of probability rather than certainty. The same thing is likely to happen in the field of pollution, leading to processes of risk evaluation that balance probabilities rather than provide firm predictions.

Such risk evaluation leads on to judgements about the justifiable degree of protection against pollution, in a world that can never be more than relatively safe, and about the liability of a polluter to one who sustains damage. As the international scale of pollution problems, and the long time span between some forms of exposure and effect, become familiar, so governments will inevitably also become involved in the evaluation of appropriate quality objectives for shared air and water, and in the assessment of liability where transfrontier pollution causes damage or the activities of the past hazard the future. There are some intractable problems here.

These problems are but some among many affecting the world environment. The population of the world continues to mount, and much of that population is inadequately fed, housed and cared for. Reserves of good agricultural land are limited and while some new areas certainly have potential, much of the food that is needed will have to come from a wiser use of land that is already managed by man. This, in turn, will not only demand more management effort (energy, fertilisers, pesticides, new genotypes and more intensive husbandry) but bring its own interactions with water quality and wildlife. If, because of inadequate money or education, this agricultural improvement does not take place, there will be even more serious problems of environmental degradation including erosion, salinisation of soil, and nutrient loss (Eckholm, 1976; Bowman, 1977). If the predictions of continuing growth of cities, especially in the developing world, are right then pollution 'hot spots' with problems of their own will tend to multiply unless forestalled by good planning and controls.

The optimism in this book – that it will be possible to control pollution even in a world with more people living at the kind of level now found in the developed countries – is only justified assuming certain standards of human sanity. The thesis is that we do now understand enough about pollution to predict and prevent its most troublesome acute manifestations, and we are increasingly recognising the less obvious risks. We are getting better at risk evaluation, and at setting realistic quality objectives and standards. Control technology has evolved as fast as, or faster than, the emergence of new hazards and there is no basic reason why better and cheaper means of pollution abatement should not be produced by human ingenuity. All this is, I believe, valid. Equally, as the density of human populations increases, as land and water are used more intensively, and as people seek a higher quality and expectation of life, we shall be able to rely less and less on the dispersal capacity of the environment and have to put more effort (and probably more money) into pollution control. The incentive to develop better abatement technology and inherently less polluting processes will be increased by these changes.

10.3 *The approach to environmental management*

Faced with these predictions and constraints, are there any general principles to guide national and local approaches to the environment? Certainly there is no universal 'cook book' recipe for managing all environments everywhere. The world is too diverse in its physical, biological and human features for this, although there do appear to be certain ingredients in common between sound recipes for planning future developments with an impact on the environment.

First, it is desirable to know in broad terms the essential features of the environmental systems involved. This calls for surveys to define not only the relief, climate, rocks, potential mineral resources and water regimes of a country or area, but the ecological systems and their potential for development or likely vulnerability to urban and industrial activities. At the same time, care is needed in designing such surveys. Not only are the parameters that could be measured legion, and the costs correspondingly great, but there is also a danger that the dynamic behaviour of environmental systems, some of which have natural cycles extending over years or decades, will not be brought out unless the surveys are repeated regularly and in fact extend into monitoring. The impact of such an extension on the costs is obvious.

The surveys of the environment have therefore to be structured to measure those components which are known to be critical either to the working of the environmental systems – those variables that determine the structure, behaviour and resilience of the whole – or are particularly relevant to types of potential development. Studies of soil nutrients, rainfall, drainage, soil–water relationships and primary production can allow prediction of what crops may grow where, with what yields and what vulnerability to drought, and what irrigation systems are needed to sustain this productivity and avoid accumulation of mineral salts. Forest surveys should define not only the best natural timber resources and the environmental causes of their quality but the areas capable of improvement (and how) and the places where forest cover must be conserved to retain soil and prevent flash floods at times of high rainfall. Other kinds of survey can define areas of highest attraction for tourists, including places rich in wildlife or important reservoirs of genetic richness, and here an understanding of ecological dynamics, population cycles, and the minimal areas to sustain viable populations will be needed. Few areas of the world are without some human impact, and before planning further development the scale of these existing disturbances, the degree of stress they are placing on systems that must be retained, and the effectiveness of the environment in dispersing their waste products, needs to be known. Many people regard this dispersal capacity as a resource and its scale certainly needs to be determined even if it is not intended to use it fully or value it financially. In fact, the kind of analysis of the pre-development situation that is needed depends on the questions to be asked: the kind of development proposed, and the type of prediction that is sought. Some kind of model of the environmental system and of the proposed development is likely to be an essential component of the analysis.

The human situation also needs survey and analysis Examinations of the pattern of human distribution and human need, of traditional boundaries between groups, of regional attitudes and goals, all need to be taken into account. The two patterns – human and environmental – both need definition, whether the evaluation is concerned with the structure plan for an English county or the development of a sparsely populated forest. In this examination, taking the social imperatives of human need, the availability of social resources, and the condition and potential of the environment together, a view can be reached about such matters as desirable environmental quality objectives and primary protection standards – and how far to use the capacity of the environment to disperse wastes. We are concerned with a two-component system: man and environment, and we must examine both.

There is rarely a unique solution to environmental management problems. Choices exist, especially in the details of the mosaic. These may express alternative food yields from different crops; alternative patterns of industrial development and their impact, or alternative blends of forest, farmland and town. Comparative analyses of this kind are well-developed skills in the planning departments of many countries, and in the international agencies. Economic values and costs must be included. Social attitudes may be decisive, including strong preferences or aversions concerning food or life styles. In poor countries where many people have a short life expectation, inadequate diet and housing, and little prospect of worthwhile employment, the imperative to development may be so strong that only the minimal standards of environmental quality needed to safeguard health and essential ecological resources are demanded. In such countries there may also be strong traditions in housing, diet and agricultural methods which constrain both development and environmental management. New patterns of land use and life style cannot be created overnight: they have to evolve, using the energies of local people, and this evolution must pass through stages all of which are socially acceptable, or are made so by education.

A margin must be allowed for error. The environment is a changing thing. It is impossible to survey and monitor all its features. Even the best predictive models are imperfect. Climatic records are broken somewhere every year. It is unwise not to allow for some margin in case the observations on which the models have been built prove to cover only a part of the actual range of variation. It is probably prudent, in developing new lands, to establish a mosaic in which areas are altered to varying degrees at least in the initial phases, so that the parts first subject to intensive management serve as natural experiments from which the wisdom of the land development strategy can be assessed. The same goes for industrial development: it is wise to allow a safety factor when calculating the attainable standards for discharges into the environment, and hold some dispersion capacity in reserve. Rather than separate the stages of survey, prediction of impact, development and monitoring of effects, therefore, it is best to make them a continuous process in which development is monitored and assessed throughout – and if necessary, modified.

An administrative system needs to be established for environmental management, and while this must, more than most things, be rooted in national and local tradition, there are certain inescapable needs. Some machinery for environmental survey, analysis, modelling and prediction must be established. A system for planning or development control is another component. One or the other may contain means for determining the capacity of the environment to receive and disperse pollutants. In the process of evaluating how changing social demands may affect the environment there must be close and continuing dialogue between the groups concerned with survey, analysis and prediction (who will be scientists) and those concerned with the administration of planning and development control (who are unlikely to be). Another essential element in the system is a machinery for monitoring changes after development, and a fourth component is a pollution control agency, ensuring that conditions imposed under planning laws are met, and taking action to impose new conditions if monitoring reveals that undesirable changes are going on. Over all these there needs to be a strategic policy body watching over the whole management of the national environment.

How the survey, planning, monitoring, control and strategic policy groups relate to ministers, local authorities and the legislature is a matter for national choice. So is the relationship between development control, control of existing pollution, agriculture and forestry, mineral exploitation and conservation of wildlife and natural beauty and amenity. Interlinkages between these groups are essential if a coherent policy is to be pursued, and the impact of all needs to be monitored so that overall policies can be adjusted. Increasingly, now, these groups also need links with their counterparts in other countries and with the international agencies.

10.4 *A final summary*

This book begins with ecology. Chapters 1 and 2 stress that 'pollution' is no more than a convenient term to describe certain physical and chemical factors whose impact in the environment has been increased by man, and whose effects are unwelcome because they damage human and other targets we value. At the end of the book, and by way of summary, ten main conclusions may be suggested.

First, there is and can be, no special science of pollution and there are dangers in treating pollutants as things apart. They must be studied using the same scientific disciplines as we apply to other physical and chemical factors that influence structures and biological systems – and with the same degree of scientific rigour.

Second, there is no special category of chemical substance we can define as 'pollutants', with criteria which allow instant recognition of any new material as belonging to the class. It is not as easy as that. Almost anything can be a pollutant if it is concentrated enough and placed where it can interact with targets we value. From this stem the two following conclusions.

Third, there is no unique classification which always serves to marshal our knowledge of pollutants and their effects: the classification we use depends on the problem we are facing and trying to solve. How we organise the information at our disposal depends on what we want it for.

Fourth, before we can analyse a pollution problem we have to get down to some hard scientific measurement. The concentration of a supposed pollutant needs to be established, at a defined location, and the pathways it may follow also have to be defined. It does not matter how toxic a substance is, or how much of it exists, if it cannot disperse to reach a target.

Fifth, we are in a very difficult field of medical and especially physiological science. The non-infectious diseases pollutants can create are not always manifest in the same way even in apparently identical targets. Like other chemical and physical factors their expression depends on the condition of the target, and that target at any time owes its condition to a multitude of interacting forces. Effects are rarely clear-cut, or instantaneous. There are almost bound to be degrees of variation and uncertainty that go ill with the emotions the subject can arouse, and with the demands for immediate and definitive action that stem from public concern.

Sixth, because of this variation and uncertainty, we are in fact faced with processes of risk estimation and with choices whose consequences we cannot fully foresee. Any standards or goals we establish can be no better than the imperfect knowledge on which we found them. They cannot be immutable, and must, indeed, be a spur to reappraisal rather than a justification for the inaction that follows the final solution of a problem.

Seventh, because of these uncertainties, we must monitor what is happening around us – how concentrations or levels of chemical and physical pollutants are changing in the environment, and how people and other targets are responding.

Eighth, we should concentrate on places where the risks are likely to be greatest (but not to the extent that we lose our capacity to detect the unforeseen). It appears probable that man is most vulnerable to chemical contaminants through his food, drink, and working environment: these, then rather than the general outdoor environment may demand greatest vigilance in monitoring and control. Equally, risks from pollution, and the costs of control, need, in a world with limited human resources, to be looked at in proportion to other hazards. So far as man is concerned there are grounds for believing that accidents in travel and at home and the ills we inflict on ourselves in various ways may be greater than the dangers of pollution.

Ninth, as we move into the future, we may expect pollution problems to be increasingly the cause of international debate and action. The carbon dioxide we release when we burn fuel is increasing in the atmosphere and could alter climate over broad regions on a significant, albeit not enormous, scale. Many substances that are widely traded may need also to be screened and monitored. No one community can act alone.

Tenth, and finally, there are grounds for optimism. There appears to be no reason why pollution should be the doom of mankind or the destroyer of the ecological balance of the world. But pollution control is an essential element in sound environmental management. The higher the standards of environmental quality and safety we demand, the more effort we shall have to make, and the more it will cost. It is a matter of wisdom and judgement how much risk is acceptable and how high environmental quality should stand amid the other goals that compete for the resources of society.

References

Advisory Committee on Pesticides and Other Toxic Chemicals (1967). *Review of the present safety arrangements for the use of toxic chemicals in agriculture and food storage.* London: HMSO.

Advisory Committee on Pesticides and Other Toxic Chemicals (1969). *A further review of certain persistent organochlorine pesticides used in Great Britain.* London: HMSO.

Allcroft, R. and Burns, K. N. (1968). Fluorosis in cattle in England and Wales: incidence and sources. *Fluoride*, **1**, 50–3.

Allee, W. C., Emerson, A. E., Park, O., Park, T. and Schmidt, K. P. (1969). *Principles of animal ecology.* Philadelphia and London: Saunders.

American Chemical Society (1969). *Cleaning our environment: the chemical basis for action*, ed. Lloyd M. Cook. Washington: American Chemical Society.

Anderson, J. M. (1971). Assessment of the effects of pollutants on physiology and behaviour. *Proc. Roy. Soc. Lond.*, **177B**, 307–20.

ARC (1967). *The effects of air pollution on plants and soil.* London: Agricultural Research Council.

Ashby, E. (chairman) (1972). *Pollution: nuisance or nemesis? A report on the control of pollution.* London: HMSO.

Ashby, E. (1973a). Prospect for pollution. *J. Roy. Soc. Arts.*, June 1973, 443–54.

Ashby, E. (1973b). Pollution across frontiers: three cases. Ernest Balsom Lecture, 1972–3. *Public Health Eng.*, March 1973.

Ashby, E. (1975). *A second look at doom.* Fawley Foundation Lecture. University of Southampton.

Ashenden, T. W. and Mansfield, T. A. (1977). Influences of wind speed on the sensitivity of ryegrass to SO_2. *J. Exp. Bot.*, **28**, 729–35.

Atkins, D. H. F., Cox, R. A. and Eggleton, A. E. J. (1972). Photochemical ozone and sulphuric acid formation in the atmosphere over southern England. *Nature, Lond.*, **235**, 372–6.

Austwick, P. K. C. (1975). Mycotoxins. *Br. Med. Bull.* **31**, 222–9.

Baker, J. M. (ed.) (1976a). *Marine ecology and oil pollution.* London: Applied Science Publishers.

Baker, J. M. (1976b). Ecological changes in Milford Haven during its history as an oil port. In *Marine ecology and oil pollution*, ed. J. M. Baker. London: Applied Science Publishers.

Baker, J. M., Campodonico, I., Guzman, L., Texera, J. J., Texera, B., Venegas, C. and Sanhueza, A. (1976). An oil spill in the Straits of Magellan. In *Marine ecology and oil pollution*, ed. J. M. Baker. London: Applied Science Publishers.

Bakir, F., Damluji, S. F., Amin-Zaki, L., Murtadha, M., Hkalidi, A., Al-Rawi, N. Y., Tikriti, S., Dhahir, H. I., Clarkson, T. W., Smith, J. C. and Doherty, R. A. (1973). Methyl mercury poisoning in Iraq. *Science*, **181**, 230–41.

Baltic Convention (1973). *Convention on protection of the marine environment of the Baltic Sea.*

Barnett, P. R. O. (1971). Some changes in intertidal sand communities due to thermal pollution. *Proc. Roy. Soc. Lond.* **177 B**, 353–64.

Beauchamp, R. S. A., Ross, F. F. and Whitehouse, J. W. (1971). The thermal enrichment of aquatic habitats. In *Proceedings of the fifth international water pollution research conference, 1970.* London: Pergamon.

Beckerman, W. (1973). Pollution control (Who should pay? How much?). In *Pollution abatement*, ed. K. M. Clayton and R. C. Chilver. Newton Abbot: David and Charles.

Beer, J. M. and Hedley, H. B. (1973). Air pollution research: reduction of combustion-generated pollution. In *Fuel and the environment. Conference proceedings*, VI. 1. London: The Institute of Fuel.

Beeton, A. M. (1965). Eutrophication of the St Lawrence Great Lakes. *Limnol. Oceanogr.*, **10**, 240–54.

Beeton, A. M. (1971). Eutrophication of the St Lawrence Great Lakes. In *Man's impact on environment*, ed. T. Detwyler, pp. 233–45. New York: McGraw-Hill.

Bell, J. N. B. (1974). Effect of ozone and peroxyacetyl nitrate on plants. In *Inter Research Council Committee on Pollution Research: Report on a Seminar in May 1974.* London: The Research Councils.

Bell, J. N. B. and Mudd, C. H. (1976). Sulphur dioxide resistance in plants: a case study of *Lolium perenne*. In *Effects of air pollutants on plants*, Society for Experimental Biology Seminar Series 1, ed. T. A. Mansfield, pp. 87–103. Cambridge University Press.

Bender, A. E. (1963). The relative merits of plant and animal proteins. In *The better use of the world's fauna for food*, ed. J. D. Ovington. London: Institute of Biology.

Bertodo, R. (1974). Pollutants from automotive sized engines. In *Inter Research Council Committee on Pollution Research*. London: The Research Councils.

Bleasdale, J. K. A. (1959). The effects of air pollution on plant growth. In *The effects of pollution on living material*, Symposium of the Institute of Biology No. 8, ed. W. B. Yapp. London: Institute of Biology.

Bourne, A. G. (ed.) (1973). *The man–food equation.* London: Academic Press.

Bowman, J. (1977). An agricultural strategy for the United Kingdom. *J. Roy. Soc. Arts.*, **128**, 365–79.

Bryce Smith, D. and Waldron, H. A. (1974). *Ecologist*, **4**, 367–77.

Bull, J. N. and Mansfield, T. A. (1974). Photosynthesis in leaves exposed to SO_2 and NO_2. *Nature, Lond.*, **250**, 443–4.

Burrows, E. M. (1971). Assessment of pollution effects by the use of algae. II. Sublethal effects and changes in ecosystems. *Proc. Roy. Soc. Lond.*, **177 B**, 295–306.

Cabinet Office (1976). *The Torrey Canyon. Report of the committee of scientists on the scientific and technological aspects of the Torrey Canyon disaster.* London: HMSO.

Calder, N. (1973). *Nature in the round.* London: Weidenfeld and Nicolson.

Capron, T. M. and Mansfield, T. A. (1977). Inhibition of net photosynthesis in tomato in air polluted with NO and NO_2. *J. Exp. Bot.* **27**, 1181–6.

Carpenter, K. E. (1928). *Life in inland waters.* London: Sidgwick and Jackson.

Catlow, J. and Thirlwall, C. G. (1977). *Environmental impact analysis in the United Kingdom.* London: Department of the Environment.

CEQ (1970, 1971a, 1972, 1973, 1974, 1975, 1976). *Environmental quality*, Annual Reports of the Council on Environmental Quality. Washington: US Government Printing Office.

CEQ (1971b). Implementation of National Environment Policy Act. Notice of opportunity for public comment on procedures. *Federal Register*, Vol. 36, No. 239, Part II, pp. 23666–712. Washington: US Government Printer.

Chadwick, M. J. and Goodman, G. T. (eds.) (1975). *The ecology of resource degradation and renewal*. 15th Symposium of the British Ecological Society. Blackwell: Oxford.

Chandler, T. J. (1965). *The climate of London*. London: Hutchinson.

Christie, A. D. (1975). Residence times in the stratosphere. In *The ecology of resource degradation and renewal*, 15th Symposium of the British Ecological Society, ed. M. J. Chadwick and G. T. Goodman. Blackwell: Oxford.

CIAP (1974a). *The effects of stratospheric pollution by aircraft*. Climatic Impact Assessment Programme Report, DOT-TST-75-50. Washington: US Department of Transportation.

CIAP (1974b). *Impact of climatic change on the biosphere*. Climatic Impact Assessment Programme Monograph 5. Washington: Department of Transport.

CIC (1975). *Environmental impact of stratospheric flight. Biological and climatic effects of aircraft emissions in the stratosphere*. Climatic Impact Committee, Washington: National Academy of Sciences.

CISC (1976). *Halocarbons. Environmental effects of chlorofluoromethane release*. Washington: Committee on Impacts of Stratospheric Change, National Academy of Sciences, National Research Council.

Clark, B. D. (1975). Project appraisal. In *Symposium on environmental evaluation*, Planning and Transportation Research Advisory Council. London: Department of the Environment.

Clark, B. D., Chapman, K., Bisset, R. and Wathern, P. (1977). *Assessment of major industrial applications*, Department of the Environment Research Report No. 13. London: Department of the Environment.

Clayton, B. E. (1975). Lead: the relation of environmental and experimental work. *Br. Med. Bull.*, **31**, 236–40.

Clayton, K. M. and Chilver, R. C. (eds.) (1973). *Pollution abatement*. Newton Abbot: David and Charles.

Clements, F. E. (1916). *Plant succession*. Washington: Carnegie Institute.

Cohen, B. L. (1977). The disposal of radioactive wastes from fission reactors. *Sci. Am.*, **236**, 21–31.

Cole, H. A. (organiser) (1971). *A discussion on biological effects of pollution in the sea*. Proc. Roy. Soc. Lond., **177 B**, 275–468.

COMESA (1975). *Report of the committee on meteorological effects of stratospheric aircraft*, Meteorological Office. London: HMSO.

Command 9322 (1954). *Report of a committee under the chairmanship of Sir Hugh Beaver*. London: HMSO.

Command 4984 (1972). *Convention for the prevention of marine pollution by dumping from ships and aircraft (The Oslo Convention)*. London: HMSO.

Command 6820 (1977). *Nuclear power and the environment. The governments response to the sixth report of the Royal Commission on environmental pollution (Cmnd 6618)*. London: HMSO.

Committee on Major Hazards (1976). *First report*. London: Health and Safety Executive.

Commoner, B. (1972). *The closing circle*. London: Cape.

Connell, J. H. and Slatyer, R. O. (1976). Mechanisms of succession in natural communities and their role in community stability and organization. *Am. Nat.*

Conway, G. R. (1976). Man versus pests. In *Theoretical ecology, principles and application*, ed. R. M. May. Oxford: Blackwell.

Conway, G. R. and Romm, J. (1972). *Ecology and resource development in South-East Asia.* New York: Ford Foundation.

Conway, V. M. (1943). The ecology of Ringinglow Moss, near Sheffield. *J. Ecol.*, **34**, 149.

Cooke, G. W. (1970). The carrying capacity of the land in the year 2000. In *The optimum population for Britain*, Institute of Biology Symposium No. 19, ed. L. R. Taylor. London: Institute of Biology.

Cooke, A. S. (1970). The effect of pp'-DDT on tadpoles of the common frog (*Rana temporaria*). *Environ. Pollut.*, **1**, 57–71.

Cooke, A. S. (1975). Selective predation by newts on frog tadpoles treated with DDT. *Nature, Lond.*, **229**, 275.

Cowell, E. B. (ed.) (1971). *The ecological effects of oil pollution on littoral communities.* London: Institute of Petroleum.

Crampton, R. F. and Charlesworth, F. A. (1975). Occurrence of natural toxins in food. *Br. Med. Bull.* **31**, 209–13.

David, O., Clark, J. and Voeller, K. (1972). Lead and hyperactivity. *Lancet*, **2**, 900–3.

Davis, J. (1971). Pollution and our environment. In *International symposium on the indentification and measurement of environmental pollutants.* Ottawa: National Research Council of Canada.

Darlington, P. J. Jr. (1957). *Zoogeography, the geographical distribution of animals.* New York: J. Wiley.

Dempster, J. P. (1975). Effects of organochlorine insecticides on animal populations. In *Organochlorine insecticides: persistent organic pollutants*, ed. F. Moriarty. London: Academic Press.

Derwent, R. G. and Stewart, H. N. M. (1973). Ozone in central London. *Nature, Lond.*, **241**, 342.

Detwyler, T. (ed.) (1971). *Man's impact on environment.* New York: McGraw-Hill.

Dicks, B. (1976). The effects of refinery effluents: the case history of a saltmarsh. In *Marine ecology and oil pollution*, ed. J. M. Baker. London: Applied Science Publishers.

DOE (1971). Department of the Environment and the Welsh Office. *Report of a river pollution survey in England and Wales, 1970.* Vols. I and II. London: HMSO.

DOE (1972a). *The human environment: the British view. Report prepared on the occasion of the United Nations conference on the human environment.* London: HMSO.

DOE (1972b). *Out of sight, out of mind. Report of a working party on sludge disposal in Liverpool Bay.* London: HMSO.

DOE (1972c). Department of the Environment and the Welsh Office. *River pollution survey in England and Wales. River quality. Updated, 1972.* London: HMSO.

DOE (1973). Department of the Environment and the Welsh Office. *Report of a survey of the discharges of foul sewage to the coastal waters of England and Wales.* London: HMSO.

DOE (1974a). *The monitoring of the environment in the United Kingdom. Pollution paper no. 1.* London: HMSO.

DOE (1974b). *Lead in the environment and its significance to man. Pollution paper no. 2.* London: HMSO.

DOE (1974c). *The non-agricultural uses of pesticides in Great Britain. Pollution paper no. 3.* London: HMSO.

DOE (1975a). *Register of research. Part IV. Environmental pollution.* London: Department of the Environment, Headquarters Library.

DOE (1975b). *Controlling pollution. Pollution paper no. 4.* London: HMSO.

DOE (1976a). *Chlorofluorocarbons and their effect on stratospheric ozone. Pollution paper no. 5.* London: HMSO.

DOE (1976b). *The separation of oil from water for North Sea operations. Pollution paper no. 6.* London: HMSO.

DOE (1976c). *Effects of airborne sulphur compounds on forests and freshwaters. Pollution paper No. 7.* London: HMSO.

DOE (1976d). *Accidental oil pollution of the sea. Pollution paper no. 8.* London: HMSO.

DOE (1976e). *Pollution control in Great Britain. How it works. Pollution paper No. 9.* London: HMSO.

DOE (1977a). *Environmental mercury and man. Pollution paper no. 10.* London: HMSO.

DOE (1977b). *Environmental standards. A description of United Kingdom practice. Pollution paper no. 11.* London: HMSO.

DOE (1977c). *Report of a survey of lead in drinking water. Pollution paper no. 12.* London: HMSO.

Doll, R. and Hill, A. B. (1964). Mortality in relation to smoking: ten years observations on British doctors. *Br. Med. J.*, **1**, 139–410 and 1460–7.

Doll, R. and others (1978). Long-term hazards to man from man-made chemicals in the environment. Report of a Royal Society Discussion Meeting. *Phil. Trans. Roy. Soc. Lond.*, **B** (in press).

Doll, R. (1978). The patterns of disease in the post-infection era: national trends. *Phil. Trans. Roy. Soc. Lond.*, **B** (in press).

Dunster, H. J. and Warner, F. B. (1970). *The disposal of noble gas fission products from the reprocessing of nuclear fuel.* AHSB (RP). R 101. London: HMSO.

Eckholm, E. P. (1976). *Losing ground. Environmental stress and world food prospects.* New York: Norton.

Egan, H. and Hubbard, A. W. (1975). Analytical surveys of food. *Br. Med. Bull.* **31**, 201–8.

Ehrlich, P. R. and Ehrlich, A. H. (1970). *Population, resources, environment. Issues in human ecology.* San Francisco: W. H. Freeman.

Elton, C. S. (1958). *The ecology of invasions by animals and plants.* London: Methuen.

Energy and the Environment (1974). Report of a Working Party. London: Royal Society of Arts.

EPA (1977). *Assessment and control of chemical problems. An approach to implementing the toxic substances control act.* Draft Paper for Discussion. Washington: US environment Protection Agency, February 17, 1977.

FAO (1971). *Pollution: an international problem for fisheries.* Rome: Food and Agriculture Organization of the United Nations.

FATE (1974). *Fuel and the environment.* Conference Proceedings, Vol. 1. London: The Institute of Fuel.

Ferry, B. W., Baddely, S. and Hawkesworth, D. C. (eds). (1973). *Air pollution and lichens.* London: Athlone Press.

Fremlin, J. H. (1964). How many people can the world support? *New Sci.*, **24**, 285–7.

Gallay, W. (1976). *Environmental pollutants: selected methods.* SCOPE Report 6. London: Butterworths Scientific Publications.

Garner, R. (1978). Prediction in the laboratory. *Phil. Trans. Roy. Soc. Lond.*, **B** (in press).

GESAMP (1970–). Group of Experts on Scientific Aspects of Marine Pollution. *Reports.* Available from the sponsoring agencies (IMCO, FAO, UNESCO, WMO and WHO).

GESAMP (1977). *Impact of oil on the marine environment,* Report No. 6. Rome: Food and Agriculture Organization of the UN.

Gibson, S. M., Lam, C. N., McCrae, W. H. and Goldberg, A. (1967). Blood lead levels in normal and mentally deficient children. *Arch. Dis. Child.,* **42**, 573–8.

Glover, H. G. (1975). Acid and ferrugineous mine drainage. In *The ecology of resource degradation and renewal.* 15th Symposium of the British Ecological Society, ed. M. J. Chadwick and G. T. Goodman. Blackwell: Oxford.

Goldberg, E. D. (1975). The mussel watch, a first step in global marine monitoring. *Mar. Poll. Bull.,* **6F**, 111.

Goldsmith, F. B. (1975). *Ecological evaluation.* Paper delivered to Conference of Institute of British Geographers, 4 January 1975.

Goldsmith, P. (1974). Pollution due to Supersonic Transport. In *Inter Research Council Committee on Pollution Research.* London: The Research Councils.

Good, R. d'O. (1953). *The geography of flowering plants.* London: Longman.

Grime, J. P. (1973). Control of species density in herbaceous vegetation. *J. Envt. Manage.,* **1**, 151–67.

Hawkes, H. A. (1974). Water quality: biological considerations. *Chem. Ind.,* **21**, 990–1000.

Hawkes, J. G. (ed.) (1978). *Environment and agriculture.* London: Duckworth.

Hawkesworth, D. L. and Rose, F. (1970). Qualitative scale for estimating sulphur dioxide air pollution in England and Wales using epiphytic lichens. *Nature, Lond.,* **227**, 145–8.

Hawkins, A. E. (1974). The electricity supply industry. In *Fuel and the environment,* Proceedings, vol. 1, pp. 55–82. London: Institute of Fuel.

Heal, O. W. and Perkins, D. (1976). IBP Studies on montane grasslands and moorlands. *Phil. Trans. Roy. Soc. Lond.,* **274B**, 295–314.

Hellawell, J. M. (1978). *Biological surveillance of rivers and lakes.* Medmenham and Stevenage: Water Research Centre.

Helliwell, D. R. (1969). Valuation of wildlife resources. *Reg. Stud.,* **3**, 41–7.

Hernberg, S. (1977). Case study: lead. Paper presented to an international conference on environmental monitoring, Bellagio, Italy, March 1977. Organised by The Rockefeller Foundation.

Hernberg, S. and Nikkanen, J. (1970). Enzyme inhibition by lead under normal urban conditions. *Lancet,* **1** (7637), 63–4.

Higgins, I. T. T. (1975). Importance of epidemiological studies relating to hazards of food and environment. *Br. Med. Bull.,* **31**, 230–5.

Higginson, J. (1975). Cancer aetiology and prevention. In *Persons at high risk of cancer. An approach to cancer etiology and control,* ed. J. F. Franemi. New York: Academic Press.

HMSO (1972). Final Act of the Intergovernmental Conference on the Convention on the Dumping of Wastes at Sea (The London Convention). London.

Holden, A. V. (1970). International co-operative study of organochlorine pesticide residues in terrestrial and aquatic wildlife, 1967–68. *Pest. Monit. J.,* **4**, 117–35.

Holdgate, M. W. (ed.) (1969a). *The seabird wreck in the Irish Sea, autumn 1969,* NERC publications, series C, no. 4. London: Natural Environment Research Council Publications.

Holdgate, M. W. (ed.) (1969b). *The seabird wreck in the Irish Sea, Autumn, 1969.* Full report, available from the Natural Environment Research Council, London.

Holdgate, M. W. (1973). Population and Pollution. In *Population and the quality of life in Britain*. London: Royal Society of Arts.

Holdgate, M. W. (1976). Ecological surveys and evaluations. In *Symposium on environmental evaluation*, PATRAC, Canterbury, 25–27 September 1975. London: Department of the Environment.

Holdgate, M. W. and Reed, L. E. (1973). The fate of pollutants. In *Fuel and the environment* (FATE), Proceedings, Vol. 1. London: Institute of Fuel.

Holdgate, M. W. and Woodman, M. J. (1975). Ecology and planning. Report of a workshop. *Bull. Br. Ecol. Soc.*, **6** (4), 5–14.

Holdgate, M. W. (1979). Targets of pollutants in the atmosphere. *Phil. Trans. Roy. Soc. Lond.*, **B** (in press).

House of Lords (1977). Session 1976/77, 22nd Report: EEC Environment Policy. R/2005/76: Water Standards for Freshwater fish. London: HMSO.

HSE (1975). *Industrial air pollution*, Health and Safety Executive, 1975. London: HMSO.

Huxley, J. S. (consultant ed.) (1973). *The Mitchell Beazley atlas of world wildlife*. London: Mitchell Beazley.

Hynes, H. B. N. (1959). The biological effects of water pollution. In *The effects of pollution on living materials*, Institute of Biology Symposium No. 8, ed. W. B. Yapp. London: Institute of Biology.

ICES (1969). *Report of a working group on pollution of the North Sea*, International Council for the Exploration of the Sea, Co-operative Research Report, Series A, No. 13, Charlottenlund, Denmark.

ICES (1970). *Report of a working group on the pollution of the Baltic Sea*. International Council for the Exploration of the Sea, Charlottenlund, Denmark.

INSTAB. Index of Solubility, Toxicity and Biodegradability of freshwater pollutants. Continually up-dated information service provided by Water Research Centre, Medmenham, Bucks. and Stevenage, Herts., UK.

Institute for Cultural Research (1975). *An eye to the future. Five views on the outlook for our environment*, ICR Monograph Series, No. 15. London: Institute for Cultural Research.

Institute of Ecology (1971). *Man in the living environment*. Report of a Workshop on Global Ecological Problems. Madison, Wisconsin: The Institute of Ecology.

IRCCOPR (1971, 1972, 1975, 1977). *Pollution research and the research councils*. Reports I–IV of the Inter-Research Councils Committee on Pollution Research. London: The Research Councils.

IRCCOPR(1974). *Some gaseous pollutants in the environment. Carbon monoxide, oxides of nitrogen, ozone, peroxyacyl nitrates, ethylene and their photochemical derivatives*, report of a seminar held at the Royal Society, May 1974. London: The Research Councils.

ITE (1976). *Annual report*. Cambridge: Institute of Terrestrial Ecology.

Jefferies, D. J. (1972). Organochlorine insecticide residues in British bats and their significance. *J. Zool., Lond.*, **166**, 245–63.

Jefferies, D. J. (1975). The role of the thyroid in the production of sublethal effects by organochlorine insecticides and polychlorinated biphenyls. In *Organochlorine insecticides: persistent organic pollutants*, ed. F. Moriarty. London, New York and San Francisco: Academic Press.

Jefferies, D. J. and Davis, B. N. K. (1968). Dynamics of dieldrin in soil, earthworms and song thrushes. *J. Wildl. Manage.*, **32**, 441–56.

Jeger, L. (chairman) (1970). *Taken for granted. Report of a working party on sewage disposal*. London: HMSO.

Johansson, O. (1959). *K. Landbr. Hogskolans Annaler.*, **25**, 57.

Jones, J. R. E. (1940). The fauna of the river Meliddwr, a lead polluted tributary of the river Rheidol in north Cardiganshire, Wales. *J. Anim. Ecol.*, **9**, 188–201.

Junge, C. E. (1972). The cycles of atmospheric gases – natural and man made. *Quart. J. Roy. Met. Soc.*, **98**, 711–29.

Kay, D. A. and Skolnikoff, E. B. (1972). *World eco-crisis. International organizations in response*. Madison, Wisconsin: University of Wisconsin Press.

Kletz, T. A. (1977). The risk equations. What risks should we run? *New Sci.* **74**, 320–42.

Kneese, A. V. (1970). The economics of environmental pollution in the United States. Paper prepared for the Atlantic Council, December 1970, and referred to by Beckerman (1973).

Kneese, A. V. (1974). The application of economic analysis to the management of water quality. Some case studies. In *The Management of Water Quality and the Environment*, ed. J. Rothenberg and I. G. Heggie. London: Macmillan.

Kneese, A. V., Rolfe, S. E. and Harned, J. W. (1971). *Managing the environment. International economic cooperation for pollution control*. New York, Washington and London: Praeger Publishers.

Kovda, V. A. (1975). *Biogeochemical cycles*. Report of a SCOPE meeting on Biogeochemical cycles, Moscow, 15–18 November 1974. Paris: SCOPE.

Lamb, H. H. (1977). *Climate: Present, past and future* (2 vols.). London: Methuen.

Lambert, P. M. and Reid, D. D. (1970). Smoking, air pollution and lung cancer. *Lancet*, **1**, 853–7.

Landsberg, H. E. (1962). City air – better or worse? In *air over cities: symposium, US Public Health Service*, Tech. Rep. A, pp. 62–5. Cincinnati, Ohio: Taft Sanitary Engineering Centre.

Langdon, F. J. (1976). Noise nuisance from surface transportation. In *Symposium on environmental evaluation*, Planning and Transportation Research Advisory Council. London: Department of the Environment.

Lansdowne, R. G. (1978). Moderately raised blood lead levels in children. *Phil. Trans. Roy. Soc. Lond.*, **B** (in press).

Lawther, P. J. (1973a). Air pollution and its effects on man. In *Pollution abatement*, ed. K. M. Clayton and R. C. Chilver. Newton Abbot: David and Charles.

Lawther, P. J. (1973b). Medical effects of air pollution. In *Fuel and the environment*. London: Institute of Fuel.

Lawther, P. J. (1975). Carbon monoxide. *Br. Med. Bull.*, **31**, 256–60.

Le Cren, E. D. and Holdgate, M. W. (eds.) (1962). *The exploi.. ion of natural animal populations*, British Ecological Society Symposium No. 2. Oxford: Blackwell.

Lee, G. F. and Veith, G. D. (1971). Effects of thermal discharges on the chemical parameters of water quality and eutrophication. In *International symposium on the indentification and measurement of environmental pollution*. Ottawa: National Research Council of Canada.

Likens, G. F., Borman, F. H., Johnson, N. M., Fisher, D. W. and Pierce, R. S. (1970). Effects of forest cutting and herbicidal treatment on nutrient budgets in the Hubbard Brook watershed ecosystem. *Ecol. Monogr.*, **40**, 23–47.

Lloyd, A. G. and Drake, J. J. P. (1975). Problems posed by essential food preservatives. *Br. Med. Bull.* **31**, 214–19.

McGinty, L. and Atherly, G. (1977). Acceptability versus democracy. *New Sci.*, **74**, 323–6.

Machta, L. (1972). Mauna Loa and global trends in air quality. *Bull. Am. Met. Soc.*, **53**, 402–20.

Machta, L. (1976). *Stratospheric ozone. An example of harm commitment.* Reports, No. 1. London: MARC, Chelsea.

MacIntyre, F. (1973). Equations of Survival. In *Nature in the round*, ed. N. Calder. London: Weidenfeld and Nicolson.

Mackenthum, K. M. (1969). *The practice of water pollution biology.* Washington: US Government Printing Office.

McKnight, A. D., Marstrand, P. K. and Sinclair, T. C. (1974). *Environmental pollution control. Technical and economic aspects.* London: George Allen and Unwin.

Maddox, J. (1972). *The doomsday syndrome.* London: Macmillan.

MAFF (1971). *Survey of mercury in food.* First report of the working party on the monitoring of foodstuffs for mercury and other heavy metals. London: Ministry of Agriculture, Fisheries and Food, HMSO.

MAFF (1973). *Survey of mercury in food. A supplementary report.* London: Ministry of Agriculture, Fisheries and Food, HMSO.

Manabe, S. and Wetherald, R. (1975). The effects of doubling the CO_2 concentration on the climate of a general circulation model. *J. Atmos. Sci.* **32**, 3–15.

Mansfield, T. A. (1974). Effects of SO_2 and NO_x on plants. In *Inter Research Council Committee on Pollution Research. Some gaseous pollutants in the environment.* London: The Research Councils.

Martin, A. E. (1975). Water supplies of the future and the recycling of drinking water. *Br. Med. Bull.*, **31**, 251–5.

Martin, P. S. and Wright, H. E. Jr. (eds.) (1967). *Pleistocene extinctions: the search for a cause.* Yale University Press.

Maugh, T. H. (1973). DDT: an unrecognized source of polychlorinated biphenyls. *Science*, **180**, 578–9.

Meadows, D. H., Meadows, D. L., Randers, J. and Behrens, W. W. (1972). *The limits to growth.* New York: Universe.

Mellanby, K. (1967). *Pesticides and pollution.* London: Collins.

Moriarty, F. (ed.) (1975a). *Organochlorine insecticides: persistent organic pollutants.* New York and London: Academic Press.

Moriarty, F. (1975b). *Pollutants and animals. A factual perspective.* London: George Allen and Unwin.

Morowitz, H. J. (1968). *Energy flow in biology.* New York and London: Academic Press.

Munn, R. E. (1973). *Global environmental monitoring systems. Action plan for phase 1.* SCOPE Report No. 3. Paris: SCOPE.

Murozumi, M., Chow, T. J. and Patterson, C. (1969). Chemical concentrations of pollutant lead aerosols, terrestrial dusts and sea salts in Greenland and Antarctic snow strata. *Geochim. Cosmochim. Acta*, **33**, 1247–94.

National Academy of Sciences (1971). *Rapid population growth: consequences and policy implications.* Baltimore: Johns Hopkins Press.

National Swedish Board of Urban Planning (undated). *Urban planning and noise from road traffic. A general guide to planners.*

National Swedish Institute for Building Research (1970). *Traffic noise in residential areas.* Stockholm.

Naylor, E. (1965a). Biological effects of a heated effluent in docks at Swansea, S. Wales. *Proc. Zool. Soc. Lond.*, **144**, 253–68.

Naylor, E. (1965b). Effects of heated effluents upon marine and estuarine organisms. *Adv. Mar. Biol.*, **3**, 63–103.

NERC (1975). *Liverpool bay.* London: Natural Environment Research Council.

NERC (1976). *Research on pollution of the natural environment: report of the ad hoc preparatory group 'F' on environmental pollution research.* Publication Series 'B', No. 15. London: Natural Environment Research Council.

NERC (1977a). Stationary Thunderstorms and severe floods. *Natural Environment Research Council News Journal*, 2(3), 12.

NERC (1977b). *Biological monitoring*. London: Natural Environment Research Council.

Nobel, P. S. (1973). Free energy, the currency of Life. In *Nature in the Round*, ed. N. Calder. London: Weidenfeld and Nicolson.

Nursall, J. R. and Gallup, D. N. (1971). The responses of the biota of Lake Wabamun, Alberta, to thermal effluent. In *International symposium on the indentification and measurement of environmental pollutants*. Ottawa: National Research Council of Canada.

OECD (1970). *Eutrophication in large lakes and impoundments*. Paris: Organization for Economic Cooperation and Development.

OECD (1972). *Survey of pollution control costs estimates made in member countries*. Paris: Organization for Economic Cooperation and Development.

Operation Oil (1970). *Clean-up of the Arrow oil-spill in Chedabucto Bay*. Report to the Ministry of Transport, Canada, and Report of the Scientific Co-ordination Team to the head of the Task Force. Ottawa: Ministry of Transport.

O'Sullivan, A. J. (1971). Ecological effects of sewage discharges in the marine environment. *Proc. Roy. Soc. Lond.*, 177 B, 331–52.

Ovington, J. D. (1963). *The better use of the world's fauna for food*. London: Institute of Biology.

PATRAC (1976). *Symposium on environmental evaluation*. Planning and Transportation Research Advisory Council. London: Department of the Environment.

PAU (1972). An economic and technical appraisal of air pollution in the United Kingdom. Chilton, Didcot, Berks.: Programme Analysis Unit.

Peachey, J. E. (1974). Environmental information and data handling. *Proc. Roy. Soc. Lond.*, 185 B, 209–19.

Pentelow, F. K. T. (1959). The general condition of the rivers of Britain. In *The effects of pollution on living material*. Institute of Biology Symposium No. 8, ed. W. B. Yapp. London: Institute of Biology.

Pentelow, F. T. K. and Butcher, R. W. (1938). Observations on the condition of the rivers Churnet and Dove in 1938. *Ann. Rep. Trent Fish. Bd.*, Nottingham.

Perkins, D. (1975). In *Institute of terrestrial ecology, annual report*. Cambridge: HMSO.

Peterson, J. T. (1971). Climate of the City. In *Man's impact on environment*, ed. T. Detwyler. New York: McGraw-Hill.

Peto, R. (1978). Detection of the risk of cancer to man. *Phil. Trans. Roy. Soc. Lond.*, B (in press).

Pochin, E. E. (1975). The acceptance of risk. *Br. Med. Bull.*, 31, 184–9.

Ponnemperuma C. (1972). *The origins of life*. London: Thames and Hudson.

Porter, E. (1973). *Pollution in four industrialised estuaries*. London: HMSO.

Ratcliffe, D. A. (1970). Changes attributable to pesticides in egg breakage frequency and eggshell thickness in some British birds. *J. Appl. Ecol.*, 7, 67–115.

Ratcliffe, D. A. (1977). *A nature conservation review*. Cambridge University Press.

Raymont, J. E. G. (1964). In *A survey of Southampton and its region*. Southampton meeting, ed. F. J. Monkhouse. London: British Association for the Advancement of Science.

Reay, J. S. (1976). UK air pollution and its control in relation to planning. In *PATRAC Symposium on Environmental Evaluation, Canterbury, September 1975*, pp. 87–96. London: Department of the Environment.

Reay, J. S. (1978). The philosophy of monitoring. *Phil. Trans. Roy. Soc. Lond.,* **B** (in press).

Regier, H. A. and Cowell, E. B. (1972). Applications of ecosystem theory, succession, diversity, stability, stress and conservation. *Biological conservation,* **4** (2), 83–8.

Rothenberg, J. and Heggie, I. G. (eds.) (1974). *The management of water quality and the environment.* London and Basingstoke: Macmillan.

Royal College of Physicians (1970). *Air pollution and health.* London: Pitman.

Royal Commission (1971). *First report of the Royal Commission on Environmental Pollution.* Command 4585. London: HMSO.

Royal Commission (1972a). *Three issues in industrial pollution. Second report of the Royal Commission on Environmental Pollution.* Command 4894. London: HMSO.

Royal Commission (1972b). *Pollution in some British estuaries and coastal waters. Third report of the Royal Commission on Environmental Pollution.* Command 5054. London: HMSO.

Royal Commission (1974). *Pollution control: Progress and problems. Fourth report of the Royal Commission on Environmental Pollution.* Command 5780. London: HMSO.

Royal Commission (1976a). *Air pollution control: An integrated approach. Fifth report of the Royal Commission on Environmental Pollution.* Command 6371. London: HMSO.

Royal Commission (1976b). *Nuclear power and the environment. Sixth report of the Royal Commission on Environmental Pollution.* Command 6618. London: HMSO.

Royal Society of Arts (1973). *Population and the quality of life in Britain.* London: Royal Society of Arts.

Ruhling, H. and Tyler, G. (1968). An ecological approach to the lead problem. *Bot. Notiser,* **121**, 321–42.

Saunders, P. J. W. (1976). *The estimation of pollution damage.* Manchester University Press.

SCEP (1970). *Man's impact on the global environment. Assessment and recommendations for action.* Cambridge, Mass.: MIT Press.

Schonbeck, H. and Van Hauf, H. (1971). Exposure of lichens for the recognition and evaluation of air pollutants. In *International symposium on the identification and measurement of environmental pollutants.* Ottawa: National Research Council of Canada.

SCOPE (1977). *Environmental issues.* SCOPE 10, ed. M. W. Holdgate and G. F. White. London and New York: J. Wiley.

SCOPE (In Preparation). *Biogeochemical cycles in the biosphere.* SCOPE 13. Paris: Scientific Committee on Problems of the Environment.

SCOPE/UNEP (1974). *Environment and development. Proceediings of SCOPE-UNEP symposium on environmental sciences in developing countries, Nairobi, February 1974.* Paris: Scientific Committee on Problems of the Environment.

Selikoff, I. J. (1978). Polybrominated biphenyls in Michigan. *Phil. Trans. Roy. Soc. Lond.,* **B** (in press).

Seppalainen, A. M., Tola, S., Hernberg, S. and Kock, B. (1975). Subclinical neuropathy at 'safe' levels of lead exposure. *Arch. Environ. Health,* **30**, 180–3.

Shaper, A. G. (1978). Cardiovascular disease and the environment. *Phil. Trans. Roy. Soc. Lond.,* **B** (in press).

Sheehy, James P., Achinger, W. C. and Simon, R. A. (1969). *Handbook of air pollution.* US Department of Health, Education and Welfare, Public Health Service Publication No. 999-AP-44. Washington: US Government Printing Office.

Sibthorp, M. M. (ed.) (1975). *The North Sea, challenge and opportunity*. David Davies Memorial Institute. London: Europa Publications.

SMIC (1971). *Study of man's impact on climate. Inadvertant climate modification*. Cambridge, Mass.: MIT Press.

Smith, J. E. (ed.) (1970). Torrey Canyon *pollution and wild life*. Cambridge University Press.

Smith, S. H. (1971). Species succession and fishery exploitation in the Great Lakes. In *Man's impact on environment*, ed. T. Detwyler. New York: McGraw-Hill.

Southwood, T. R. E. (1972). The environmental complaint: its causes, prognosis and treatment. *Biologist*, **19**, 85–94.

Southwood, T. R. E. (1976). Bionomic strategies and population parameters. In *Theoretical ecology: principles and applications*, ed. R. M. May. Oxford: Blackwell.

Spicer, A. (1975). Toxicological assessment of new foods. *Br. Med. Bull.*, **31**, 220–1.

Standing Technical Committee on Synthetic Detergents (1957–77). *Progress reports*, 1–17. London: HMSO.

Sullivan, F. M. (1978). Congenital malformations and other reproductive hazards. *Phil. Trans. Roy. Soc. Lond.*, **B** (in press).

SVEAG (1976). *Oil Terminal at Sullom Voe. Environmental impact assessment*. Sullom Voe Environmental Advisory Group Report. Sandwick, Shetland: Thuleprint.

Svensson, B. H. and Soderlund, R. (eds.) (1976). *Nitrogen, phosphorus and sulphur: global cycles*. SCOPE 7. Ecological Bulletins, 22. Stockholm: Swedish Natural Science Research Council.

Sweden (1972). *Air pollution across national boundaries. Impact on the environment of sulphur in air and precipitation. Sweden's case study to the United Nations Conference on the Human Environment*. Stockholm: Royal Ministry of Foreign Affairs/Royal Ministry of Agriculture.

Tansley, A. G. (1935). The use and mis-use of vegetational terms and concepts. *Ecology*, **16**.

Taylor, G. R. (1970). *The doomsday book*. London: Thames and Hudson.

Tucker, A. (1972). *The toxic metals*. London: Earth Island.

Tukey, J. (ed.) (1965). *Restoring the quality of our environment*. Report of the Environmental Pollution Panel, President's Science Advisory Council. Washington: US Government Printing Office.

UNEP (1976a). *The state of the environment, 1976*. Nairobi: United Nations Environment Programme.

UNEP (1976b). *A conference of plenipotentiaries of coastal states of the Mediterranean region for the protection of the Mediterranean Sea*. Geneva: United Nations Office of Public Information.

UNEP (1977). *The state of the world environment, 1977*. Nairobi: United Nations Environment Programme.

US Department of Health, Education and Welfare (1968). *Progress in the prevention and control of air pollution*. Washington: Senate Document No. 92; 90th Congress, 2nd Session, June 28, 1968.

Uzawa, H. (1972). Optimum management of Social overhead capital. In *The Management of water quality and the environment*, ed. J. Rothenberg and I. G. Heggie. London and Basingstoke: Macmillan.

Valentine, D. H. (ed.) (1972). *Taxonomy, phytogeography and evolution*. New York and London: Academic Press.

Vogel, F. (1978). Our load of mutation: reappraisal of an old problem. *Phil. Trans. Roy. Soc. Lond.*, **B** (in press).

Wainwright, S. J. and Woolhouse, H. W. (1975). Physiological mechanisms of heavy metal tolerance in plants. In *The ecology of resource degradation and renewal*, 15th Symposium of the British Ecological Society, ed. M. J. Chadwick and G. T. Goodman. Oxford: Blackwell.

Ward, B. and Dubos, R. (1972). *Only one earth: the care and maintenance of a small planet*. London: Andre Deutsch.

Wardley Smith, J. (1976). Oil spills from tankers. In *Marine ecology and oil pollution*, ed. J. M. Baker. London: Applied Science Publishers.

Warren Spring Laboratory (1972). *National survey of air pollution, 1961–1971*, Vols. 1–3. London: HMSO.

Weiss, I. and Lamb. H. H. (1970). On climatic change in the North Sea, and the outlook. *Fachliche Mitteilungen*, no. 160, May 1970, 6–15.

Welsh Office (1975). *Report of a collaborative study on certain elements in air, soil, plants, animals and man in the Swansea, Neath and Port Talbot area together with a report on a moss bag study of atmospheric pollution in South Wales*. Cardiff: Welsh Office.

Weston, R. L., Gadgil, P. D., Salter, B. R. and Goodman, G. T. (1965). Problems of revegetation in the Lower Swansea Valley, an area of extensive industrial dereliction. In *Ecology and the industrial society*, ed. G. T. Goodman, R. W. Edwards and J. M. Lambert. Oxford: Blackwell.

Wheeler, A. (1969). Fish life and pollution in the lower Thames. A review and preliminary report. *Biol. Conserv.* 2 (1), 25–30.

Wiliams, R. J. H. and Ricks, G. R. (1975). Effects of combinations of atmospheric pollutants upon vegetation. In *The ecology of resource degradation and renewal*, 15th Symposium of the British Ecological Society, ed. M. J. Chadwick and G. T. Goodman. Oxford: Blackwell.

WMO (1975). *Proceedings of World Meteorological Organization/ International Association of Meteorology and Atmospheric Physics Symposium on long-term climatic fluctuations*. WMO Publication No. 421. Geneva.

Wolff, J. (1978). Pollution of the Rhine. In *The restoration of damaged ecosystems in temperate lands*, ed. M. W. Holdgate and M. J. Woodman. New York: Plenum Press.

Woodiwiss, F. S. (1964). A biological system of stream classification. *Chem. Ind.*, 14 March 1964.

Yapp, W. B. and Watson, D. J. (eds.) (1958). *The biological productivity of Britain*. London: Institute of Biology.

Yapp, W. B. (ed.) (1959). *The effects of pollution on living material*, Institute of Biology Symposium No. 8. London: Institute of Biology.

Index

abortion 125
absorption of pollutants 41, 52, 57, 59, 60, 62, 63, 130, 144
accidents 40, 141–3
accumulator organisms 28, 39, 62, 177, 186
acetylcholine 121
'acid rain' 49, 51, 134, 236
adaptation 4–6, 62, 113
administration of pollution control 206, 207, 211–19, 255
Advisory Committee on Pesticides etc. (UK) 104, 216
aerosols 51, 75
agencies for pollution control 211–19
agriculture
    damage to 40, 50, 71, 77, 85, 109, 130–4, 150, 192, 192
    development of 7, 13, 14, 243
    pollution from 95, 108, 128, 129
Agung, Mount 73
air, see atmosphere
aircraft, pollution from 48, 73–7, 116, 117
ALA-D 118–20, 146, 147
aldrin 29, 87, 104
algae 63, 89, 95, 139–3, 136, 137
Alkali and Clean Air Inspectorate (UK) 151, 160, 161, 171, 183, 199, 206, 207, 213
Alkali etc. Works Regulation Act (UK) 206, 207, 249
aluminium smelters 133
amenity 195, 196
Ames Test 37
ammonia 78, 87, 131
ammonium sulphate 78, 87
Antarctic 57, 77, 86, 87, 170
Arrow 97
arsenic 28
asbestos 59
atmosphere
    effects of pollution on 66–8, 69–88
    origin and structure of 2, 46, 48, 86
    pollutant dispersion in 44, 46–52, 84, 86, 103, 130, 223
    pollutant reactions in 26, 27, 39, 49–51
    pollutants in 25–7, 34, 66–88, 133–5, 192, 193

    radiation transmission in 69–71, 73–8
    monitoring of 166–74, 177, 179–84
    standards and pollution controls 80–3, 149, 150, 152, 153, 213
auks 127, *see also* guillemots

bacteria 31, 37, 38, 53, 55, 90, 136, 137
Baltic Convention 226, 237
Baltic Sea 54, 65, 101, 180, 224
baseline surveys 175
baseline stations 180
bats 123
Beaver Committee (UK) 80, 192
benefits
    environmental 196–7
    and pollution 190, 191, 196–7, 234
benzene hexachloride 29, 87
beryllium 125
'best practicable means' 88, 152, 160, 161, 206, 213
bioaccumulation 28, 29, 35, 36, 45, 53, 54, 62, 63, 91, 147, 186
biochemical effects of pollution 37, 110, 113, 114, 118–25, 137, 146
biodegradation 28, 29, 45, 137
biogeochemical cycles 2, 3, 13, 17, 29, 53, 65, 70, 71, 106, 107, 179, 195
biogeographical changes, 77, 118
biological indicators 132, 134
Biological oxygen demand (BOD) 90
biomes 3, 77
birds
    effects on 63, 104, 121, 122, 125–9, 136, 137
    residues in 12, 57, 61, 62, 65, 87, 102, 127, 177, 185, 186
'black-lists' 29, 30, 34
blood, pollutants in 57, 59–62, 83, 84, 118–20
bone 61, 62
boundary layer 46, 48, 50
bronchitis 82, 244

cadmium 28, 54, 62, 91, 120
calcium 13, 122
California 84, 85
Canada 122, 180, 204, 223, 236, 237, 247
canals 89

271

cancer, carcinogenesis 21, 31, 37, 42, 56, 74,
    77, 96, 116, 117, 124, 125, 137, 244
carbon cycle 2, 38, 52, 69–73, 84, 106, 179
carbon dioxide 52, 69–73, 76, 224
carbon monoxide 55, 83, 84, 86, 150
carbonates 70, 71
carbophenothion 104
cardiovascular disease 21, 83, 84, 120
Central Unit on Environmental Pollution
    (UK) 180, 188, 189, 212
charges for pollution 201
chemical factors 5, 16, 19, 54, 65, 110, 166–74
chimney height and dispersion 82, 160
chlor-alkali works 101
chlorine 51
chlorofluorocarbons 51–3, 74–7, 106, 116
Ciona 62
citrus 50
Cladophora 136
classification of pollutants 22–5, 27, 42
Clean Air Acts (UK) 80–3
Clear Lake (USA) 63
climate, effects of pollution on 67–73, 75–8,
    222, 227
cloudiness 73
Clyde, River 102
coal 79, 87
codes of practice 161
combustion, pollution from 23, 25–7, 69, 70–5,
    78–84, 107, 233
coniferous trees 132, 150
control of pollution 35, 39, 40, 42, 81–3, 86,
    88, 90, 103–5, 202
    administration of 203, 206, 211–19, 231–40
    costs of 143, 190–2, 197–201, 234, 248
Control of Pollution Act (UK) 103, 206, 215,
    216
copper 120
corals 137
cosmetics 21, 41
costs
    of pollution damage 109, 190–5, 197, 234
    of pollution control 143, 190–2, 197–201,
    234, 248
Council on Environmental Quality (USA)
    65, 193, 202, 211, 212, 213
Council of Europe 227, 237
'criteria', *see* dose:effect relationships
crops, *see* agriculture
cyclamates 42
'cyprinid' rivers 153, 157

damage
    acute and chronic 108, 109, 115, 143,
    244
    costs 109, 190–5, 197, 234
Danube, River 88, 223

Daphnia 63
data banks 42, 43
DDD, DDE, DDT
    accumulation of 60–3, 87
    effects of 38, 63, 118, 120–4, 129
    in environment 20, 29, 38, 53, 57, 63, 65,
    87, 224
    controls on 87, 144
decomposers 6, 7, 8, 118, 136
'Delany clause' (USA) 42
Department of Environment (UK) 22, 171,
    184, 212
depots, in body 60–2
derived working limits/levels 148
deserts 77, 78
desulphurisation 78, 79
detergents 28, 29, 41, 94, 162, 215, 216
development 1, 2, 13–15, 242
    control 103, 204–9
dieldrin 29, 63, 87, 104, 122–4, 128
diesel smoke 83, 160
dioxins 21, 125
dispersion of pollutants
    in air 35, 36, 38, 44, 48, 50–2, 57, 149, 173,
    248, 249
    in water 35, 38, 44, 53, 57, 90–7, 149
diversity, ecological 11, 139
DNA 116, 124
dose 36, 64, 140, 147, 148
    commitment 64, 76, 77, 140
    :effect relationships 34, 64, 111, 113–25,
    140, 147, 148, 169, 179
dumping
    at sea 17, 100–3, 161, 178, 216, 217, 237
    on land 103, 104, 152, 161, 216

'early warning' systems 50
earthworms 62, 63
'ecodevelopment' 2
economic
    basis for controls 151, 191, 192, 197–201
    effects 109
    evaluations 151, 190–201
Economic Commission for Europe (ECE) 31,
    160, 227
ecological factors 4, 5, 19
ecosystems
    effects of pollution on 16, 78, 106, 110, 111,
    113, 118, 128, 129, 134–9, 169
    impact of man on 12–16, 67, 71, 73, 78
    pathways in 58, 110, 111
    response to stress 11–13, 138, 139
    structure of 3, 6–9, 11, 111
effects of pollution
    in air 66, 67–88, 222, 227
    on animal physiology 37, 110, 113, 114, 121–5
    on animal populations 113, 118, 124–9

effects of pollution (*cont.*)
 on biochemical systems 37, 110, 113, 114,
  118–21, 124, 125
 on climate 71–8
 on ecosystems 16, 69, 78, 106, 109–11, 113,
  118, 128, 129, 134–9, 129
 factors influencing 114, 115, 116, 122, 123,
  125, 127, 128, 141–3, 250
 on land 103, 104
 on man 79–83, 109, 116, 117, 241–7
 on plants 50, 109, 129–35
 types of 20, 21, 25, 30, 36, 37, 108–16, 131, 250
 in water 66, 69, 88–103, 105, 117, 118
eggs, residues in 87, 177
eggshells, effects on, 121, 122, 128, 129
Ekofisk (oilfield) 99
electrostatic precipitators 88, 160, 171
Eliminius modestus 118
elimination of pollutants 60–3, 113, 123, 124
emphysema 84
energy
 and agriculture 14, 15
 flow in ecosystems 6, 7
 generation 15, 17, 19, 52, 67–9, 73, 78, 79,
  87, 105, 117, 233
 and pollution 18, 19, 67, 70, 71, 73, 78–82,
  84, 233
environment, *see also* sectors
 capacity of 97, 150–60, 227–31, 235, 236,
  248, 249
 management of 164, 169, 252–5
 pollutants in 25–31, 44, 45, 64–107
 surveys of 163, 173–5, 204, 205, 252, 253
 values of 155, 162, 163, 195, 196
environmental impact assessment 202–9
environmental quality objectives, standards,
 see objectives, standards
Environment Protection Agency (USA) 42,
 212
enzymes 118–21, 123, 136, 137
epidemiology 37, 245, 250
estimated dose (ED) 36
estuaries, pollution of 53, 54, 89, 91–5, 101,
 102, 118, 133, 137, 138, 152, 153
Europe, pollution in 84, 101–3, 106
European Economic Community (EEC) 42,
 153, 157, 160, 180, 215, 226, 232, 234, 237
eutrophication 12, 88–90,. 94, 118, 135–6
evolution 3, 132
excretion 60–3
exploitation of populations 12, 14, 129
exposure to pollution 52, 57, 64, 115, 130, 140,
 143–50, 172

Factory Inspectorate, (UK) 149, 213
fat reserves, pollutants in, 61, 62, 122, 123,
 127, 128, 177

feathers
 effects on 122, 125, 127, 128
 pollutants in 61, 62
fertilisers
 use of 13–15, 161
 pollution by 75, 76, 89, 107, 116
fire, impact of 13
fish
 effects on 89, 91, 102, 118, 122, 129, 135–9,
  153, 155
 pollutants in, 53, 55, 57, 62, 63, 101, 127,
  137
fisheries, pollution and 102, 109, 118, 137, 153,
 157, 194, 195
fluidised bed combustion 79
fluorine, fluoride 51, 55, 130, 131, 133
food
 additives 21, 41, 42, 56
 monitoring 172, 186, 247
 pollution of 21, 31, 40, 56, 63, 95, 137, 144,
  148, 171, 244, 247
 production 7, 9, 12, 13–15
food chains 6, 7, 38, 62, 63, 127
forests, clearance of, 13, 14, 71, 73, 89
fresh waters
 changes in 88–97, 135
 eutrophication of 88–90, 135, 136
 monitoring 184, 185
 pathways and pollutants in 27–9, 34, 44,
  52–6, 63, 88–97, 135, 136, 160
 standards and controls 96, 152–7, 159, 169,
  215
 supply 55, 56, 95, 96, 97, 120, 147, 154, 155,
  215, 246
frog (tadpole), effects on 122, 123
fungi 31, 56, 134

geese 104
genetics and pollution 113, 116, 117, 132
GESAMP 17, 29–33, 237
Glasgow 81
Global Atmospheric Research Programme
 (GARP) 179, 180
global effects of pollution 65, 66, 68–73, 76,
 83, 105, 106, 107, 222, 248
 monitoring 106, 179–81
 standards 233
Global Environmental Monitoring System
 (GEMS) 180, 181, 239
goals, *see* objectives
Great Ouse, River (UK) 104
Great Lakes, N. America
 changes in 65, 88, 194, 223, 249
 Water Quality Agreement 180, 223, 229, 237
Greater London Council (UK) 81, 82
'greenhouse effect' 69, 71–3, 75–7
'grey-list' of pollutants 29, 34

gross national product 199, 234
guillemot (*Uria aalge*) 122, 125–8
gulls 122

haem, haemoglobin 62, 118, 119
Halobates 62
harm commitment 64, 76, 77, 140
Hawaii 70, 73
health and pollution 152, 153, 155, 158, 194, 244, 245
Health and Safety Executive, UK 207, 213
heat as pollutant 19, 66–9, 117, 118
  in cities 66–8, 117
  in water 69, 117, 118
herbicides 96, 104
herons (*Ardea cinerea*) 104, 128, 187
hexachlorophene 21, 125
homeostasis 111, 113, 119, 124, 169
'hot spots' of pollution 66–9, 87, 101, 106, 117, 223, 248
hydrocarbons 74, 83, 84, 86, 99, *see also* oil
hydrochloric acid 51, 133
hypertension 120

ice caps 77
impacts, *see* effects
indicators 132, 178, 185, 186, 187
industry
  exposure in 34, 41, 117, 125, 140–3, 171, 244
  pollution from 26, 27, 39, 41, 48, 67, 68, 69, 87–91, 94, 96, 97, 99–101, 103, 117, 133, 183
  standards and controls 42, 87, 88, 97, 140, 144, 146, 148, 149, 151, 155–63, 171, 213, 214, 242
  stress from 15, 16, 125
Industrial Pollution Inspectorate (Scotland) 183
industrial revolution 86, 134
insects, effects on 59, 121–4, 129
insecticides, *see* pesticides
INSTAB 28
Institute of Terrestrial Ecology (UK) 186
Intergovernmental Maritime Consultative Organisation (IMCO) 178, 237
International Atomic Energy Agency (IAEA) 105
International Commission on Radiological Protection (ICRP) 161, 225
International Council for the Exploration of the Seas (ICES) 102, 180, 226
International Council of Scientific Unions (ICSU) 227
International Registry of Potentially Toxic Chemicals (IRPTC) 240
International pollution
  action on 238–40

conventions on 226, 229, 235–8
  liabilities for 221, 251
  standards 231–3
  trends in 250–1
  types of 78, 220–5
Iraq 244
Ireland 126, 127
Irish Sea 54, 101, 102, 105, 125–8, 224

Japan 244

kidney 60–2, 120
krypton 52

Laminaria 132, 133, 137
Lancashire 132
land
  pollution of 34, 103–5
  pollution control 103, 104, 216
laws, on pollution 144, 155, 162, 163, 233, 235
Law of the Sea 236
LC_{50}, LD_{50} 36
lead
  in body 55, 57, 59, 61, 119, 120, 144, 146, 172
  effects of 118–21, 125, 249
  in environment 20, 28, 48, 55, 57, 84–6, 120, 246
  in industry 87, 88, 120, 209
  standards for 86, 146, 150, 171, 194, 246
Lebistes 63
Lee, River (UK) 94, 95
liability, international 219, 222, 235
lichens 130, 132–5, 150, 176, 177, 187, 188
life strategies ('*r*' and '*K*') 9, 11–14, 16, 37, 139
liver 62, 123, 124, 127, 128
livestock, pollution and 31, 40, 109
'load on top' 231
local government and pollution control 103, 213, 215, 216
London 50, 65, 67, 79–84, 89, 194, 223, 245, 246, 249
London Convention 31, 103, 178, 216, 237
Los Angeles 50, 66, 84, 132, 223, 249
lung, uptake through 57, 59

Maclaren, Lake (Canada) 65
man
  development and impact of 1, 12, 13, 16, 55, 56, 106, 107, 243, 247
  effects on 18, 19, 31, 50–7, 109, 119, 120, 125, 140, 243–7
  health of 125, 146–9, 152, 172, 193, 194, 232, 244, 245
  population of 241–9
Manchester 81, 89

marine ecosystems, effects on 132, 133, 136, 137, *see also* sea
Mauna Loa, Hawaii 70, 73
medicine 18, 21
Mediterranean 180, 226, 237
melanoma 117
mercury
    effects in man 40, 63, 120, 125, 244, 249
    in fish 53, 55, 63, 101, 224, 232, 245
    pollution by 20, 28, 38, 40, 53, 55, 56, 62, 63, 96, 100, 101, 172
Mersey, River (UK) 91
metals
    in environment 19, 28, 29, 84, 91, 100, 102, 103, 113, 186, 224
    smelting 19, 103, 187, 247
methaemoglobinaemia 95, 96, 147
methane 89
Metula 97
Michigan 40
Milford Haven 137
Minamata 65, 125, 249
mining 28
ministries, and pollution control 211–19
Ministry of Agriculture, Fisheries and Food (UK) 171, 186
molluscs 54, 62, 91
models 45, 51, 57, 59, 71, 73, 74, 110, 113, 114, 179, 204, 205, 233
    of world trends 241–3
monitoring
    biological 177, 178, 185–8
    in Britain 102, 180–9
    definitions 164, 166, 170, 171
    factor 166–74, 180–5
    international 170, 178, 179, 226
    methods 170–83
    purposes 164, 166, 189
    target 166–78, 185–8
Monitoring Assessment Research Centre (MARC) 164, 226
Monitoring Management Groups (UK) 188
Morecambe Bay (UK) 188
moss 85, 177
muscles, pollutants in 62, 127, 128
'mussel watch' 39
mutagenesis 117, 124, 125
mycotoxins 31, 56, 244

National Academy of Sciences (USA) 74
National Environmental Policy Act (USA) 151, 202, 228
Natural Environment Research Council (UK) 221
nervous system, effects on 61, 62, 119–22, 125
Netherlands 102
Nesocichla eremita 87

New York 67
nitrate
    and eutrophication 12, 89, 90, 136
    in water 56, 89, 90, 95, 96, 106, 147
nitric acid 135
nitrite 56
nitrogen cycle 2, 106, 107
nitrogen oxides 74–7, 84, 107, 116, 130, 131, 135, 150
nitrosamines 31, 56, 96
noise
    controls 117, 215
    as pollutant 19, 117
    standards 209, 210, 211
Norfolk Broads (UK) 90
North Atlantic Treaty Organisation (NATO) 227
North Sea 54, 99, 101, 102, 105, 224
Norway 49, 50, 134
nuclear power 18, 52, 105
nutrients, plant 6, 13, 19, 20, 107, 135, 136

objectives, environmental quality 81–3, 144, 150–60, 162, 163
    international 228–33
oil
    controls 99, 103, 217, 231, 236, 237
    effects 102, 103, 136, 137
    in fresh waters 26, 53, 96
    at sea 29, 30, 53, 97–9, 102, 136, 137, 224
    sources 57, 99, 102
oil refineries 102, 103, 137, 187
organic wastes 28, 89, 90, 136
Organization for Economic Co-operation and Development (OECD) 31, 179, 199, 223, 226, 232, 234
organochlorine compounds 20, 29, 39, 61–5, 87, 100, 102, 104, 120–5, 128, 129, 186, 224, 232
organophosphorus compounds 29, 36, 121
Orkney Islands 103
Oslo Convention 17, 31, 102, 103, 180, 216, 226, 237
Oslo Fiord 54
oxidants 84, 85, 132, 133, 147, 150
'oxidant smog' 39, 84, 85
oxygen
    cycle 2, 38, 106
    in fresh waters 54, 65, 88–92, 136
    in inshore seas 55, 65, 101, 102
ozone
    as pollutant 84, 130–3, 150
    stratospheric 48, 51, 74–6, 116, 224

Panonychus ulmi 129
Paris 83
Paris Convention 17, 31, 99, 103, 151, 237

particulates 59, 69, 73, 76, 84, 87, 88, 130, 133
pastoralists, impact of 14
pathways of pollution 17, 44–63, 110
  in air 44–52, 87, 130, 149
  in ecosystems 57, 58
  in food chains 56, 62, 63, 247
  in fresh waters 44, 45, 52–4, 149
  modified by man 55, 56
  in organisms 27–62, 87
  in sea 44, 45, 52–7, 87, 97, 127, 128, 149
  in soil 55, 56, 139, 149
penguins 57, 87
Pennsylvania 84
persistence of pollutants 20, 29, 35–8, 51, 52,
  75, 87, 101, 129
peroxyacetyl nitrates (PAN) 26, 39
pesticides 14, 15, 20, 28, 29, 38, 63, 84, 87, 100,
  104, 120–5, 128, 129, 161
Pesticides Safety Precautions Scheme
  (UK) 104
petrol 48, 86
pharmaceuticals 21, 28, 42, 96
phosphate
  eutrophication by 12, 89, 90, 94, 136
  sources of 94
phosphorus cycle 2, 106, 107
photochemical reactions 26, 38, 39, 84, 132
photosynthesis 2, 6, 7, 38, 71, 130, 131
physical factors 4, 64–7, 110, 166, 174
  properties of pollutants 22, 23, 45
physiological effects
  in animals 113–15, 119, 120–5
  in plants 130, 131
pigeons 63
plankton 38, 62, 63, 127, 136
plants, exposure to pollution 52, 57, 59,
  130
  effects on 50, 109, 113, 129–36
planning 202–7
'polluter pays principle' 199–201, 234
pollutants, pollution, *see also* under effects,
  pathways, environmental media and
  particular properties or aspects
  classification of 22–27, 42
  context of 1–16
  controls of 35, 39, 40, 81–3, 103–5, 199–219,
    231–40
  cost of 109, 143, 190–201, 234, 248
  definition 17, 18, 21
  prediction of 40–3
  properties of 19–21, 35–41, 44, 49, 51, 62,
    69–71, 75, 84, 87, 113–16, 244
  sources of 20, 23, 24, 26–9, 32, 33, 78, 82–90,
    99, 102, 103
  standards for 143–63
  types of 19–35
polybrominated biphenyls (PBBs) 21, 40, 56

polychlorinated biphenyls (PCBs) 29, 40, 57,
  62, 65, 100, 102, 104, 106, 122, 123, 125–8,
  224
populations
  dynamics 128, 129
  effects on 125–9, 137, 139
Porphyra 105, 172
power stations 67, 69, 79, 87, 105, 117, 118
predators 14, 62, 63, 104, 128, 187
pregnancy 125, 140
productivity, biological 7–9, 14–16, 77
Programme Analysis Unit 192, 193

radioactive materials, wastes 18, 52, 105, 216,
  217, 224
radiation
  as pollutant 18, 74, 76, 105, 116, 117, 124,
    144
  through atmosphere 69–78
Ranunculus fluitans 28, 177
rats 61
reactions of pollutants 35, 38, 39, 49, 51, 69–71,
  84, 131, 244
receptors, *see* targets
recycling 152
red spider mite 129
reservoirs, pollution in 89, 95
residues in tissues 51, 61, 62, 63, 87, 102, 104,
  122, 123, 147, 185, 186
'reference man' 140, 245
resilience 11, 110, 127, 128
Rhine, Convention and Commission 227, 237
Rhine, River 65, 88, 102, 223
risk
  estimation 17, 40, 108, 140–3, 150, 151, 162,
    163, 257
  extent of 106, 120, 140–3, 256, 257
rivers, *see* fresh waters and by name
Rome 84
Rotterdam 84
Royal Commission on Environmental Pollution (UK) 65, 69, 94, 152, 153, 213, 217
Royal Commission on Sewage Disposal
  (UK) 90, 96
ruthenium 105, 172
ryegrass 130–3

saccharin 42
safety margins 41, 120, 142, 143, 152, 245, 247
salmonid fishes 122, 129, 135, 136, 153, 155
'salmonid' rivers 153, 157
sampling, of environment 172–5
Scientific Committee on Problems of the
  Environment (SCOPE) 164, 170, 179,
  226
Scotland 127
screening, of substances 36, 37–42

sea
control of pollution 151, 178, 216, 217, 236
dumping of wastes in 17, 100, 101, 178, 216,
    217, 237
effects on 69, 117, 118
level changes 73, 77
oil pollution of 28, 97–9, 213
pathways in 44, 53–7, 62, 71
pollution of 27–33, 40, 54, 69–71, 97–193,
    105, 118, 125–8, 132–3
seabirds
oil and 97
pollution and 40, 65, 87, 102, 122, 125–8
Seveso 65
sewage
as pollutant 88–91, 95, 102, 118, 136
treatment 89–90, 92, 94, 96–7
shearwaters 87
Sheffield 134
shellfish 54, 91, 137
Shetland 103, 204
ships, pollution from 97, 99, 237
'sinks' for pollution 29, 44, 49, 51, 53–5, 70,
    71, 73, 99, 100, 149, 221, 228, 235
skin
effects on 116, 117
uptake through 52, 59
smells 83, 126
smoke control areas 80, 81
smoke
emissions 50, 78–84, 245
effects 50, 79, 81, 84
monitoring 183, 184
standards 149, 152, 184
social costs 195, 197–201
social values 162, 163, 195, 196
socio-economic context of pollution 108–10,
    142, 151, 162–6, 190–201, 253–5
soil
erosion of 13, 14
fertility 13, 14
nutrients 13, 14, 19, 107, 136
pollution 19, 38, 55, 103–6, 109, 113, 129,
    130, 134
sound as pollutant 19, 117
sources of pollution 26, 27, 44–9, 82, 83–90,
    99, 100, 102, 103
South Pole 70
Southampton (UK) 118
SSTs *see* aircraft
stability, ecological 11, 19
standards
ambient 148
biological 147
environmental quality 81–3, 90, 95, 148–55
emission 158–60
international 156, 158–60, 231–6

primary protection 144–8, 152, 232
process 160–1
product 161–2, 230
revision of 162, 163, 201
types of 143–63
uniform 158–60
vehicle 160
'standard man' 140, 245
Standing Committee on Synthetic Deter-
    gents (UK) 215
starlings 67
Stockholm 1, 17, 83, 227, 237, 238
stomata 59, 130
storm drainage 89, 90
strategies, r and K 9, 11–14, 16, 37, 139
stratosphere 46, 48, 51, 52, 74–6
stress
and ecosystems 11–13, 15, 138, 139
and pollution 122, 125, 127, 128, 139
types of 12, 13, 15, 16, 28
sub-lethal effects 114, 115, 121, 122, 129, 148
succession, ecological 8–11
Sullom Voe 103, 204
sulphur
cycle 2, 47, 49, 78, 79, 106, 107, 179
in fuels 79
sulphur dioxide
control 79–83
dispersion 49, 79, 83, 134, 223, 235
effects 79–81, 84, 130–5
emissions 78, 79, 81–3, 169, 245
monitoring 173, 174, 183, 184
standards 81, 149, 152, 153, 169, 174, 184
'sulphuretum' 118, 136
sulphuric acid 49, 134, 135
sunbathing 74, 117
sunshine and pollution 81
surveys 164, 173–6
surveillance, *see* monitoring
supersonic aircraft, see aircraft
suspended solids 90
Sweden 49, 65, 85, 134, 209, 211
synergism 131

tankers, *see* ships
targets (or receptors)
effects on 25, 108–39
exposure of 52, 57, 140–9
importance of 108, 190, 191
monitoring of 166–70
nature of 17, 108–15
primary and secondary 110, 112
sensitivity of 122, 123, 127, 129–34, 140, 141,
    146
Tees, River (UK) 91
'Teesside mist' 78, 87
teleconnections 77

temperature and pollution 67, 71, 73, 118, 122, 123
teratogenesis 21, 31, 124, 125
tests of pollutants 12, 36–8, 41, 42, 115
textile industry 94
Thames, River (UK) 89, 91, 92, 95, 102
thresholds and pollution 140, 142
Threshold Limit Values (TLVs) 36, 140, 149
thrushes 63
thyroid 122
tobacco 132
Torrey Canyon 97, 137
TOVALOP 231
Toxic Substances Control Act (USA) 41
toxicity
  acute 36, 37, 115, 147
  chronic 36, 37, 115, 148
  factors affecting 19, 20, 36, 37, 122, 123, 125, 127, 128, 130–2
  of pollutants 19, 35–7, 113–16
  testing 21, 115
trade, barriers to 222, 223, 231, 232
Traill smelter 235
transfrontier pollution 220–2, 230
Trent, River (UK) 69, 91
Trent Biological Index 91
trends in pollution 64–107
Tristan da Cunha 87
trithion 104
trophosphere 46, 48, 50, 75, 76, 86
tuna 232
Tyne, River (UK) 91

ultra-violet radiation 38, 51, 74, 116, 117
Ulva 132, 187
United Kingdom
  air pollution in 49, 50, 79–83
  control of pollution in 80–3, 103–6, 159, 160, 199, 212–17
  land pollution in 103–5
  monitoring in 102 180–9
  planning in 203–5
  pollutant trends in 65, 79–84, 88–106, 160
  sea pollution in 101–5
  water pollution in 88–97
United Nations Conference on the Human Environment 1, 17, 225, 226, 227, 237, 238
United Nations Economic Commission for Europe (ECE) 31, 160, 227
United Nations Food and Agriculture Organisation (FAO) 227
United Nations Educational, Scientific and Cultural Organisation (UNESCO) 227

United Nations Environment Programme (UNEP) 17, 31, 42, 178, 180, 225, 226, 239
United Nations Scientific Committee on Effects of Atomic Radiation (UNSCEAR) 105, 225, 232
United States
  air pollution in 84, 85, 152, 153
  control of pollution in 104, 106, 152, 153, 159, 160, 180, 199, 202, 232
  monitoring in 174
  water pollution in 88, 95, 159
urban pollution 67, 68, 78–83, 84, 87, 88, 119, 120, 160, 209, 223, 245, 246
Uria aalge, see guillemot
Urothoe brevicornis 118
UV (B), see ultra-violet

values, value judgements 108, 109, 153, 155, 162, 163, 190, 191, 193, 201, 234
vanadium 62, 120
variability, in target response 115, 116, 122, 123, 130–2, 140, 141, 146, 172
vegetation, surveys of 175, 176
Verrucaria 97, 187
volcanic emissions 47, 49

Warren Spring Laboratory (UK) 50, 185
waste
  disposal on land 18, 103, 104, 152, 161, 216
  disposal at sea 17, 18, 100–3, 152, 161, 217, 237
waste heat 66–9, 117, 118
water, see fresh waters and sea
water supplies 55, 56, 95–7, 120, 147, 154, 155, 215, 246
Water Authorities (UK) 151, 171, 183, 184, 206
weather and pollution 67, 68, 71, 81, 127, 130, 131
wild life
  monitoring 185–7
  value of 109
Windscale (UK) 105
World Health Organisation (WHO) 31, 81, 95, 96, 149–55, 173, 178, 184, 194, 227, 232, 246
World Meteorological Organisation (WMO) 69, 71, 78, 79, 226, 233
'wreck' of seabirds 125–8

X-rays 18, 144

zinc 103, 120